Advanced Series in Agricultural Sciences 2

Harry Wheeler

Plant Pathogenesis

With 19 Figures

Springer-Verlag
Berlin Heidelberg New York 1975

Harry Wheeler, M. S., Ph. D., Professor of Plant Pathology, University of Kentucky, Lexington, USA

ISBN 3-540-07358-2 Springer-Verlag Berlin · Heidelberg · New York
ISBN 0-387-07358-2 Springer-Verlag New York · Heidelberg · Berlin

Library of Congress Cataloging in Publication Data. Wheeler, Harry, 1919 — Plant pathogenesis. (Advanced series in agricultural sciences; 2) Bibliography: p. Includes index. 1. Plant diseases. I. Title. II. Series. SB731.W495 581.2 75-19318

Typesetting, printing and binding: Brühlsche Universitätsdruckerei, Gießen.

Foreword

The *Advanced Series in Agricultural Sciences* is designed to fill a long-felt need for advanced educational and technological books in the agricultural sciences. These texts, intended primarily for students of agriculture, should also provide up-to-date technical background reading for the many agricultural workers in extension services, educational systems, or international bodies.

The editors of *Advanced Series in Agricultural Sciences* will select key subjects relating to the agricultural environment, agricultural physics and chemistry, soil science, plant sciences, animal sciences, food technology, and agricultural engineering for a critical and synthetic appraisal. An initial theoretical presentation will be used by authors of individual volumes in the series to develop a technical approach—including examples and practical solutions— to each subject.

In addressing the advanced undergraduate and early graduate student of agriculture, selected authors will present the latest information, leavened with the lessons learned from their own experience, on precise and well-defined topics. Such books that widen the horizons of the student of agriculture can serve, too, as useful reference sources for the young specialist in the early years of his career.

Many specialists who are involved in teaching agricultural science are isolated from universities and research institutions. This series will bring them up-to-date scientific information, thus keeping them in touch with progress.

The basic objective of *Advanced Series in Agricultural Sciences* is to effect a structural integration of the theoretic and technical approaches to agriculture. The books will be particularly helpful to extension specialists who have an ever-present need for the latest information in the day-to-day solving of practical problems.

The increasing involvement of agricultural sciences in projects in developing countries has created a demand for clear, current texts on specific problems. The texts to be published in this new series, written by specialists from different countries, should provide this profession with the appropriate tools for insuring the effectiveness of schemes for agricultural development all over the world.

The normal activities of the editors of *Advanced Series in Agricultural Sciences* center around teaching and research, not publishing. It was our awareness of the scarcity of advanced texts that led us to accept Springer-Verlag's invitation to enter this field and to devote time to writing. We hope our endeavors will be met by the understanding of our colleagues—both scientists and teachers—and we hope that they will cooperate with us by using *Advanced Series in Agricultural Sciences* in the way we envisage, so making a useful contribution to agriculture.

The Editors

Preface

This book attempts to meet the objectives of the editors of this series by providing concise, critical, synthetic, and up-to-date coverage of a selected subject in the field of agricultural sciences. The subject chosen, plant pathogenesis, is obviously too broad to be covered comprehensively in a brief text. In selecting material to be included, emphasis has been placed on recent developments in rapidly expanding areas of research, an approach which has its hazards, as exciting new ideas and concepts often have very short half-lives.

The book is intended to serve as a text for advanced undergraduate and graduate students in plant sciences and as a reference source for students, researchers, and technologists.

I thank all those who granted permission to reproduce illustrations or figures: acknowledgment for these is given in the text. I am much indebted to colleagues who have reviewed various chapters: Richard Chapman, Stephen Diachun, Vernon Gracen, Jr., Joseph Kuć, Thomas Pirone, Charles Poneleit, Carroll Rawn, and especially to one who read the entire book: Penelope Hanchey. My thanks also go to Mrs. Ellen Elbel for preparing line drawings and to Mrs. Debby Owen for typing the manuscript.

January 1975 HARRY WHEELER

Contents

Chapter 1 Concepts and Definitions

A. Plant Disease

Oral inquisitions, required of those who aspire to advanced degrees in plant pathology, often begin with a request for the candidate's definition or concept of plant disease. After a struggle, the candidate extracts from a brain freshly crammed with scientific names of pathogens, disease cycles, biochemical pathways, and a variety of other disorganized information, a definition memorized in some introductory course. Usually this definition attempts to distinguish normal or healthy plants from abnormal or diseased ones. This leads to a discussion of the meaning of the term "normal" and to the conclusion that, by the candidate's definition, plants which produce five times the average yield for a given locality must be diseased.

To escape the trap, the candidate modifies the definition to specify that diseases represent undesirable abnormalities. Now virus-induced floral variegations, some of which are highly prized, are brought to the candidate's attention. Eventually, after also failing to distinguish clearly in all cases disease from injury, the candidate concedes that he is incapable of formulating a satisfactory definition of plant disease. He consoles himself with a mental note that his examiners are probably equally incompetent.

However defined, plant diseases have two important aspects: one economic, the other ecologic. On occasion, one or both of these aspects of a plant disease can be spectacular. Famine and mass human migration caused by potato blight in Ireland, virtual elimination of native chestnuts from North American forests by chestnut blight, and the destruction of more than 10 million acres of maize in a single year (1970) by leaf blight are a few well-known examples of the profound economic and ecological effects of plant diseases. Less known but more important are the smaller, often unrecognized economic losses and ecological changes caused by a multitude of diseases which constantly attack wild and cultivated plants. Quite naturally, plant pathologists have emphasized the economic aspects of plant diseases on major crops and natural resources rather than the more subtle ecological impacts.

Broadly conceived, plant diseases include all malfunctions which result in unsatisfactory plant performance or which reduce a plant's ability to survive and maintain its ecological niche. Pathogenesis, the subject of this book, is the sequence of events which occur during disease development.

B. Pathogens and Parasites

Traditionally, two general classes of agents which cause plant diseases have been recognized. One class consists of germs; the other of agents unlike germs. A variety of terms have been used for the two classes—parasitic and nonparasitic—infectious and noninfectious—animate and inanimate—biotic and abiotic. The term pathogen clearly applies to classical biotic agents (fungi, bacteria, and viruses) which cause most plant diseases. Some also consider abiotic physical and chemical agents to be pathogens, but most avoid this term, especially when deficiency of an essential material is responsible for disease symptoms.

Plant disorders caused by animals, other than nematodes, are generally excluded from the area of plant pathology. Genetic abnormalities, of great importance in human pathology, are rarely mentioned as causes of disease in plants. This is not because they do not occur but rather because they are constantly eliminated through artificial and natural selection. Pathogenic higher plants, though few in number, cause several diseases of major importance. Dwarf mistletoes which threaten coniferous forests provide one example.

During the past decade two new types of plant pathogens have been described. One, discovered in Japan, is a wall-less, membrane-bound, mycoplasma-like organism. It causes diseases of the yellows type which had long been attributed to viruses (Davis and Whitcomb, 1971). The second, a subviral agent termed a viroid, has been implicated as the cause of potato spindle tuber, chrysanthemum stunt, and citrus exocortis (Diener, 1972). Viroids are RNA molecules which lack a protective protein coat. How they are replicated in host cells is a major mystery since they are theoretically too small (ca. 100000 daltons) to carry the necessary genetic information for their own replication.

Many biotic pathogens are also parasites, that is, they not only cause disease, they also live with and obtain food from the plants they attack. For such associations the term pathogenism, defined as a parasitic, symbiotic relationship in which one symbiont causes disease in the other, has been proposed (Hall, 1974). Hall is among those who feel the term symbiosis has been widely misused in biology by restricting it to associations in which both members benefit (mutualism) or in which one benefits without harmful effects on the other (commensalism). In this view, Hall joins Henry (1966), Bracker and Littlefield (1973) and others who regard parasitism, mutualism, and commensalism as different types of symbiosis. In the absence of a consensus, the term symbiosis will be avoided and the term parasite used for an organism or virus which exists in intimate association with another living organism from which it derives an essential part of the material for its existence (Thrower, 1966).

Although we are concerned with pathogenesis rather than parasitism, the extent to which a pathogen has acquired the parasitic habit must be considered as an important factor in disease development. Traditionally, biotic plant pathogens have been classified as obligate parasites (unable to exist apart from their hosts), facultative saprophytes (parasitic habit strongly developed), facultative parasites (parasitic habit weakly developed), or saprophytes (incapable of parasitism). The fact that a number of fungi once considered to be obligate parasites have been cultured apart from their hosts on artificial media makes the term obligate inappropriate. The term facultative is both cumbersome and confusing. Comprehension of the phrase "facultatively saprophytic" requires a mental pause for translation to "strongly parasitic."

In a discussion of parasitism as an aspect of fungal ecology which influences concepts of classification and evolution, Luttrell (1974) recognizes three broad categories.

I. Biotrophs	—organisms which, regardless of the ease with which they can be cultured, in nature obtain their food from living tissues on which they complete their life cycles. Typified by rusts, smuts, and mildews.
II. Hemibiotrophs	—organisms which attack living tissues in the same way as biotrophs but continue to develop and sporulate after the tissue is dead. Typified by leaf-spotting fungi.
III. Perthotrophs	—organisms which kill host tissues in advance of penetration and then live saprophytically. Typified by fungi such as *Sclerotium rolfsii* which utilize sclerotia or other food bases in the course of initial attack.

This classification, to which subcategories can be added, should be useful where formal terms are required. Informally, parasites can be simply termed strong or weak.

C. Pathogenicity and Disease Reaction

Pathogenicity, the ability to cause disease, can be considered an attribute of a particular individual agent or as a characterisitc of a group of agents. Currently, the trend is toward the latter usage with the term virulence being used for the disease-inducing capacity of an individual member of the group. This convention makes possible reference to avirulent isolates of a pathogenic species or to pathogens of wheat avirulent to barley (Wood, 1967). Adoption of this convention requires caution. Obvious self-contradictions, such as "avirulent pathogen," should be avoided.

Plant reactions to pathogens can be described as varying degrees of either resistance or susceptibility. Often disease ratings, ranging from 0 (no symptoms) to 4 or some higher value which denotes maximum symptom development, are employed. For practical and certain theoretical purposes, plants rated below some arbitrary value (usually 2 or 3) are considered resistant whereas those with higher ratings are designated susceptible.

The intensity of cellular response to a pathogen is often inversely related to the severity of disease suffered by the whole plant. Reactions in which a limited number of host cells are quickly killed and the pathogen contained in a localized necrotic lesion are termed hypersensitive. Hypersensitivity can be viewed as extreme cellular reactivity which may confer a high degree of disease resistance to the whole plant. Since hypersensitive reactions reflect the inability of the plant and pathogen to co-exist, these are often termed incompatible. Reactions in which the pathogen ramifies within the plant without rapid killing of host cells are termed compatible. The latter, of course, usually result in whole-plant susceptibility.

The concept of tolerance as a type of resistance to plant disease has been discussed by Schafer (1971). Tolerant plants have the ability to survive and perform satisfactorily at levels of infection that cause unacceptable losses to other plants of the same species. Unlike hypersensitivity, tolerance does not depend on the ability to limit the growth and development of the pathogen. Certain diseases caused by viruses and nematodes provide good examples of tolerance. In tolerant plants, the virus multiplies and spreads rapidly but causes only mild symptoms. Similarly, tolerant plants continue to make satisfactory growth while supporting the growth and development of large numbers of nematodes.

In the Western Hemisphere, the term immunity is commonly applied when plants fail to develop visible symptoms when exposed to a pathogen under conditions favorable for disease development. Some, however, define immunity as absolute freedom from disease (Nelson, 1973). Under this definition, immunity would be ruled out by any pathological response of a plant to a pathogen no matter how minute or transient the response might be. If such immunity exists, it must be very rare. Even *Ginkgo biloba*, famed for its freedom from disease, responds to the presence of potential pathogens which are incapable of penetrating even its cuticle (Adams et al., 1962). In the USSR and Eastern Europe, immunity is used in quite a different sense. Scientists in these countries employ the term plant immunity where Western scientists would use plant disease resistance (Rubin and Artsikhovskaya, 1963). In view of these different usages, the term should be avoided or defined when used.

D. Problems of Terminology

Only a few of the more common terms employed by plant pathologists have been discussed in this Chapter. Despite repeated attempts, precise, universally acceptable definitions for many of these terms have not been formulated. The same is true for many other terms which will be introduced in later chapters.

One problem is the difficulty of coining concise expressions which will accurately and completely describe all aspects of complex phenomena. For example, a mature maize plant with two or three minute leaf spots caused by a frustrated potential pathogen would not be considered diseased from the economic or ecologic viewpoints discussed in the first section of this chapter. However, to a plant pathologist struggling to understand the nature of disease, the cells in these spots are definitely diseased. Thus, plant diseases have a third aspect albeit one that is essentially academic.

Other terminology problems stem largely from lack of understanding of the phenomena on which the terms are based. Completely satisfactory terms and definitions must await complete understanding of the fundamental nature of plant disease. Until then most definitions should be considered as tentative and subject to modification as new information is gained.

Chapter 2 Mechanisms of Pathogenesis

Pathogenesis can be viewed as a battle between a plant and a pathogen refereed by the environment. A small change in a single environmental factor can decide the outcome of the plant-pathogen struggle. For example, lines of wheat which carry a gene known as Sr 6 are highly resistant to race 56 of stem rust when grown at 20° C. If the temperature is increased to 25° C, the plants become highly susceptible to the same race of this pathogen (Daly, 1972). Single biological components of the environment can be equally decisive. Damage to tomato plants caused by *Sclerotium rolfsii* can be greatly reduced by adding a second fungus, *Trichoderma harzianum*, to the soil. In this case, biological control of the disease has been attributed to the ability of *Trichoderma* isolates to overgrow and kill *S. rolfsii* (Wells et al., 1972).

In nature, environmental referees, chiefly climatic conditions, physical and chemical properties of soils, and the plant surface microflora, are constantly interacting with each other and with the plant and pathogen to determine the course of pathogenesis. Matta (1971) points out that such complex interactions can be studied better in the laboratory, where some of the components can be controlled, than in nature where they cannot. In controlled environments we can focus attention on the activities of the two combatants, the plant and pathogen. In so doing, we must keep in mind that the events observed under controlled, hence artificial, conditions may not resemble those which occur in nature. Furthermore, even rigid control of physical and chemical factors does not eliminate, and may in fact enhance, the activities of the rich variety of microorganisms, mostly nonpathogens, which colonize plant surfaces.

In soil-borne diseases, the importance of microorganisms of the rhizoplane (root surface) and the rhizosphere (area immediately surrounding the root) has long been recognized (Garrett, 1956). More recently, corresponding aerial regions, the phylloplane and phyllosphere, have received attention (Preece and Dickinson, 1971). A discussion of the roles which these microorganisms play in nature in predisposing or protecting plants and in interacting synergistically or antagonistically with pathogens is beyond the scope of this book. Instead we can emphasize only the hazards which these microorganisms introduce into the conduct and interpretation of laboratory experiments. Most studies of pathogenesis involve inoculation of plants or plant parts under conditions highly favorable for the growth of bacteria and other microbes. Anyone who has attempted to reisolate *Colletotrichum* from artificially infected plants has encountered the "yellow peril," a well-known but rarely mentioned pigmented bacterium which is constantly associated with lesions produced by fungal pathogens of this genus.

While concentrating on the plant and pathogen, we must not ignore the omnipresent saprobic third component of the system.

A. Initial Stages of Pathogenesis

I. Contact

Pathogenesis caused by infectious agents begins when the pathogen contacts an infectible region of the plant. Most fungal and many bacterial pathogens make contact with plants fortuitously in the form of wind—or water-borne spores or propagules. Most viruses, many bacteria, and some fungi are carried to plants by insects or other vectors. Evidence that mobile pathogens are attracted to plants by root exudates (chemotaxis) has been summarized for zoospores by Hickman and Ho (1966) and for nematodes by Webster (1969). They conclude that although chemotaxis is a common phenomenon there is little evidence for specific attraction to susceptible roots.

Electrotaxis has been the subject of conflicting reports for both zoospores and nematodes. In a study of seven species of *Phytophthora*, Khew and Zentmyer (1974) observed three types of electrotactic responses. At currents of less than 0.5 μA, zoospores were attracted to the anode. At higher current intensities, a repulsion zone occurred at the anode and accumulation, caused by immobilization and trapping rather than attraction, occurred at the cathode. They concluded that current intensities within the range around plant roots (0.3–0.6 μA) are capable of initiating electrotaxis and suggested that the different types of responses observed might account for previous conflicting results.

II. Entry

1. Through Wounds and Natural Openings

Bacterial pathogens gain access to the plant interior either through wounds or through stomates, hydathodes, lenticels, or nectaries. In general, bacteria do not invade living cells; instead they colonize intercellular spaces and nonliving elements of the vascular system. Infection of roots of legumes by nitrogen-fixing bacteria (*Rhizobium* spp.) provides an exception to the rule. Here an infection thread penetrates the host cell walls and the bacteria contained within the thread are then released into the host protoplasm (Dixon, 1969). Although this mode of entry would appear to convey an advantage, it has not evolved among the bacteria now recognized as plant pathogens.

Many fungal pathogens also enter plants only through wounds or natural openings. Two general types can be recognized. Some, often called "wound pathogens," enter wounds and grow only in dead or dying plant tissues. The second category is typified by most rust fungi. These enter through stomates, grow intercellularly and produce specialized structures called haustoria which penetrate the host cell wall and invaginate the protoplast. Haustorial penetration will be discussed with other types of direct penetration in the following section.

Viruses, which are known to multiply only inside living protoplasts, enter intact plants only through wounds; usually those caused by a vector. Most vectors are insects, but nematodes (Hewitt et al., 1958), mites (Slykhuis, 1969) and even primitive fungi (Campbell and Grogan, 1963) have been shown to transmit viruses. Work with mechanically wounded plants suggests that a transitory, reparable type of wound is required. Damaged leaf hair cells and exposed channels in epidermal walls (ectodesmata) have been postulated to be avenues by which virus particles gain entry (Matthews, 1970).

Some viruses are capable of infecting isolated protoplasts (cells from which the walls have been removed enzymatically). Electron micrographs indicate that virus particles may enter isolated protoplasts through vesicular invaginations of the protoplast surface—a process called endocytosis (Cocking, 1970). If so, this would suggest that the transitory wound required for infection would be merely one that briefly exposed an area of the protoplast surface. However, the possibility that ruptures in the protoplast surface brought about during isolation are required for virus infection of protoplasts cannot be ruled out. Polyornithine, an agent known to damage cells and disrupt permeability, enhances viral infection of isolated protoplasts (Takebe and Otsuki, 1969).

2. Direct Penetration

Once contact has been made, direct penetration by a pathogen involves the following steps: the spore or propagule itself, or a germ tube produced by it, or, most often, a specialized structure (appressorium) produced by the germ tube becomes firmly cemented to the plant surface; a penetration peg then forms which passes through the cuticle, if present, then through the cell wall, and into the cell lumen. Two questions regarding the penetration process will be considered. First, is the penetration peg forced through the cuticle and cell wall mechanically by growth or osmotic pressure, or is its passage facilitated by enzymatic degradation of these barriers? Second, to what extent do these barriers to penetration confer disease resistance?

a) Penetration of the Cuticle

Aerial plant parts are covered by a cuticular membrane composed chiefly of waxes, cutin, and lipids. Cutin, a complex polymer of fatty and hydroxy-fatty acids, is the chief structural component. Waxes occur embedded in the membrane and extruded in a marvelous variety of forms on its surface. Cellulosic and pectic components predominate where the membrane merges with the epidermal cell walls.

Improved techniques in scanning and transmission electron microscopy have added a new dimension to our knowledge of the anatomy and structure of leaf surfaces. These same techniques combined with light microscopy and histochemistry are currently being employed to study penetration by plant pathogens. A volume edited by Preece and Dickinson (1971) provides an introduction to these techniques and a summary of our knowledge of the structure and chemistry of cuticles as they relate to the activities of pathogenic and nonpathogenic microorganisms which colonize leaf surfaces.

Detailed studies of cuticular penetration have been carried out with powdery mildews (*Erysiphe* spp.), anthracnoses (*Colletotrichum* spp.) and the soft-rot pathogen *Botrytis cinerea.* Scanning electron micrographs of germinated spores of *E. graminis* f. sp. *hordei* on its host, barley (Fig. 1a) and on a nonhost, cucumber (Fig. 1b) illustrate the ability of this technique to reveal details of surface structure. The spores and germ tubes can be removed by coating the plant surfaces with molten gelatin then removing the gelatin layer after it has hardened. When this is done, traces left by the fungus on the plant surface can be seen (Fig. 1c, d). On both plants, penetration holes are round with smooth edges and show no signs of physical stress or tearing. Moreover, on barley the surface wax crystals appear dissolved in areas beneath germ tubes and hyphae (Fig. 1c). These observations suggest that enzymatic activity is involved in cuticular penetration. The evidence, however, is not conclusive. Staub et al. (1974) point out that fungal secretions probably contribute to the imprint images in Fig. 1c, d and such secretions or a fungal sheath (Fig. 1e, f) could adhere to the leaf and mask the wax crystals. Hyphae of *Helminthosporium maydis* ramify extensively beneath the cuticle prior to penetrating the cell wall. Hence, it is possible that structures which appear to be fungal sheaths in Fig. 1e, f actually represent the ruptured host cuticle.

Transmission electron micrographs of sections through penetration pegs of *Colletotrichum graminicola* on maize (Figs. 2, 4) are typical of those obtained with many plant-pathogen combinations. These show the cuticle distinctly depressed inward as the penetration peg enters the cell wall. Such inward displacement indicates that mechanical forces play some role in cuticular penetration by such pathogens. On the other hand, results with broad bean *(Vicia faba)* leaves inoculated with *Botrytis cinerea* suggest that mechanical force may not be required. In this case, no evidence of inward depression of the cuticle during penetration was found (Fig. 3). Furthermore, esterase activity was detected histochemically at germ tube tips at the time of penetration but was not detected 16 h later. Similar transitory esterase activity has been reported for *Venturia inaequalis*, a fungus which ramifies between the cuticular membrane and the cell wall (Nicholson et al., 1972). McKeen (1974) suggests that the transitory nature of enzyme activity may account for previous failures to extract cutin-degrading enzymes from *B. cinerea.*

b) Penetration of Cell Walls

In general, evidence from electron microscopy indicates that enzymatic activity plays an important role in cell-wall penetration. The inward depression of the cuticle which is often observed (Figs. 2, 4) could hardly occur if the underlying layers of the cell wall had not undergone softening or partial degradation. Furthermore, swelling (Fig. 3) or changes in staining properties (Figs. 4a, 5b) of the

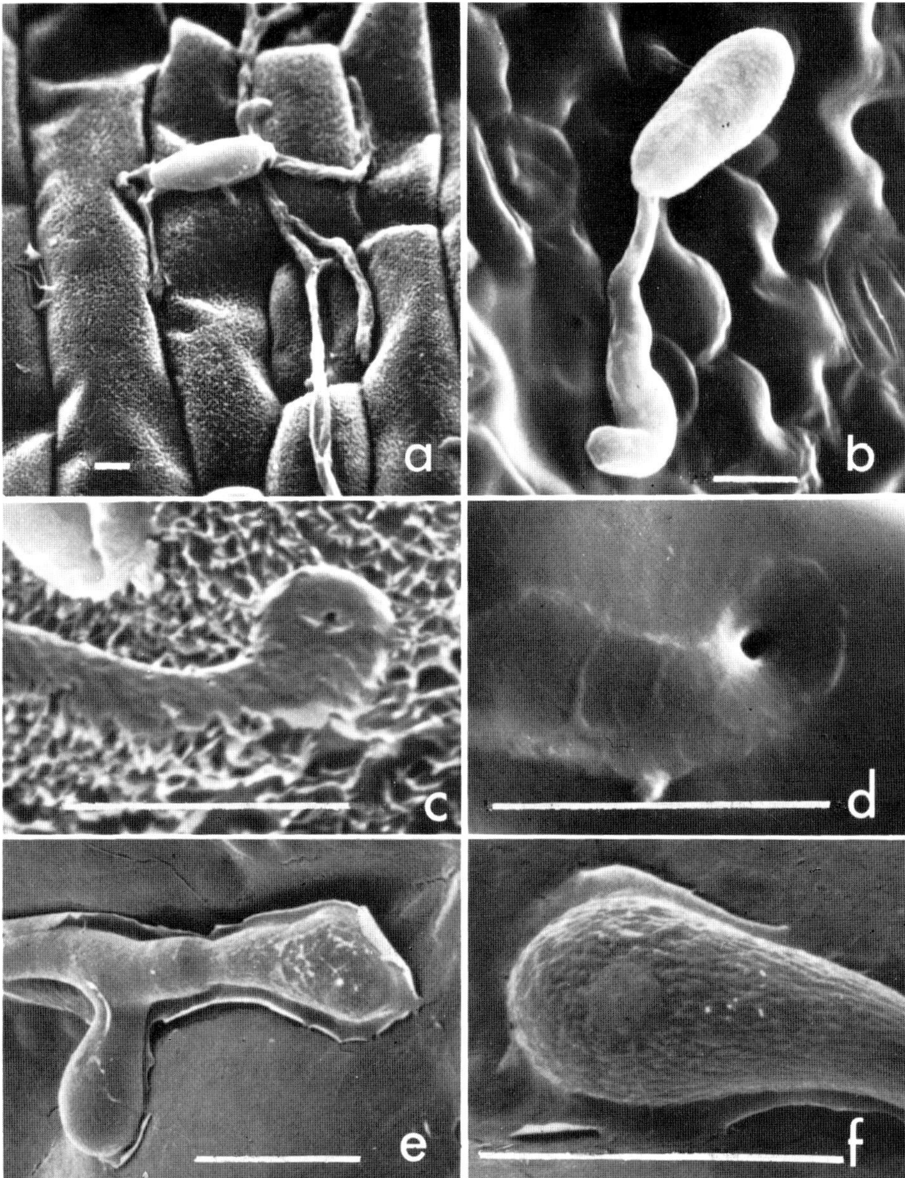

Fig. 1a—f. Scanning electron micrographs of *Erysiphe graminis* spores and germ tubes on barley leaves (a) and on cucumber cotyledons (b). Imprints and perforations seen on barley (c) and on cucumber (d) after *E. graminis* was removed. (e) and (f) structures which resemble sheaths on hyphae of *Helminthosporium maydis* growing on maize leaves may represent the ruptured host cuticle. Bar equals 10 μm. (a)–(d) reproduced by permission from Staub et al. (1974). (e)–(f) Courtesy of R. O. Blanchard

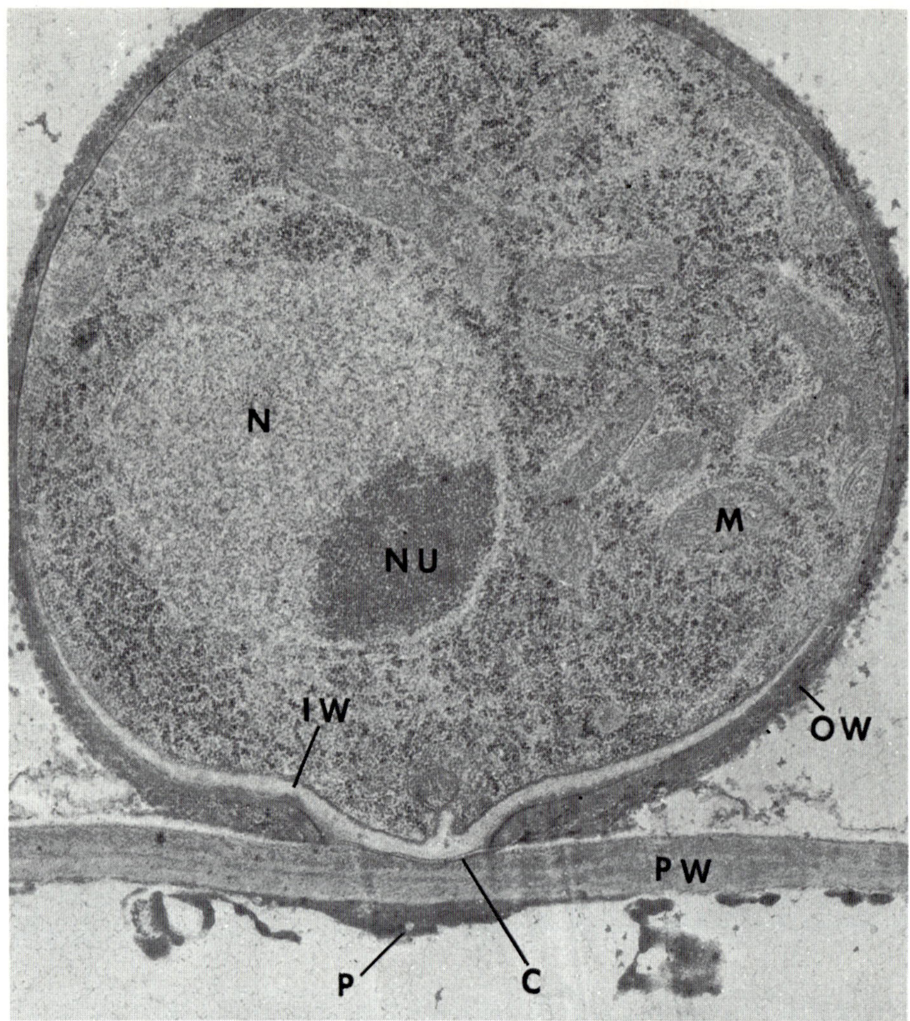

Fig. 2. Initial stage of penetration of an epidermal cell wall *PW* of maize by *Colletotrichum graminicola* 6 hrs after inoculation. The mature appressorium, surrounded by a dense outer wall *OW*, contains the usual organelles; mitochondria *M*, a nucleus *N*, a nucleolus *NU* and groundplasm rich in ribosomes. A penetration peg formed by an electron-lucent inner wall *IW* in the floor of the appressorium is in contact with the thin cuticle *C* which is slightly depressed inward. A small papilla *P* has formed on the inner surface of the plant cell wall (×24000). Courtesy of D. J. Politis

host cell wall often occur in advance of the penetration peg or hypha. Finally, localized invaginations similar to those produced by penetration pegs of *Colletotrichum lindemuthianum* in futile attempts to pierce films of epoxy resins (Fig. 4c) have not been observed at any stage of cell wall penetration.

Enzymatic activity during cell-wall penetration by fungi such as *Erysiphe* and *Colletotrichum* appears sharply localized. (Figs. 1, 2, 4). This suggests that the en-

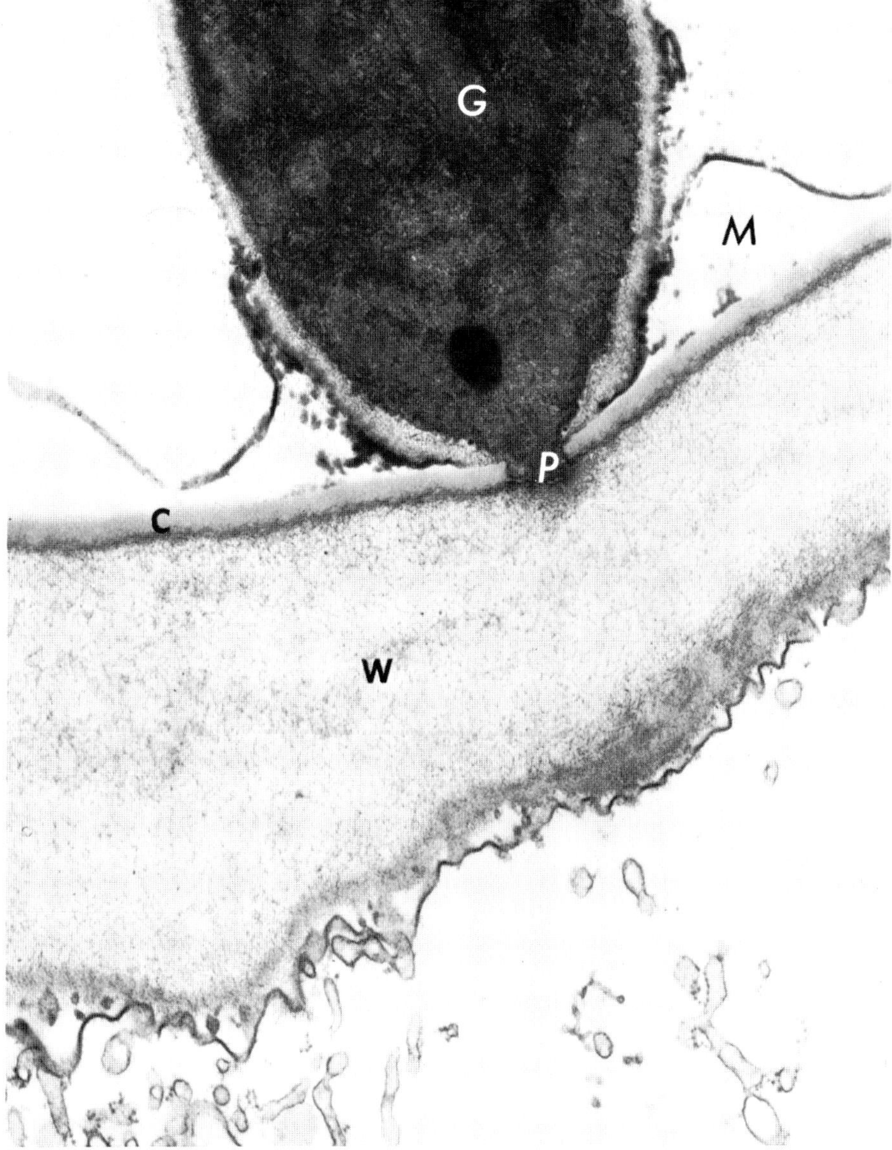

Fig. 3. Infection of *Vicia faba* by a germ tube *G* of *Botrytis cinerea*. The infection peg *P* has penetrated the cuticle *C* and has entered the swollen cell wall *W*. Mucilage *M* is present on both sides of the germ tube (× 32600). Reproduced by permission from McKeen (1974)

zymes involved may be bound to the tip of the penetration peg or that some material produced only at the tip of the peg causes a transient activation of enzymes bound to host cell-wall components. In either case we are faced with the problem of how an enzyme or an enzyme-activator can be localized at the exterior surface of the cell wall which usually surrounds the penetration peg of the pathogen.

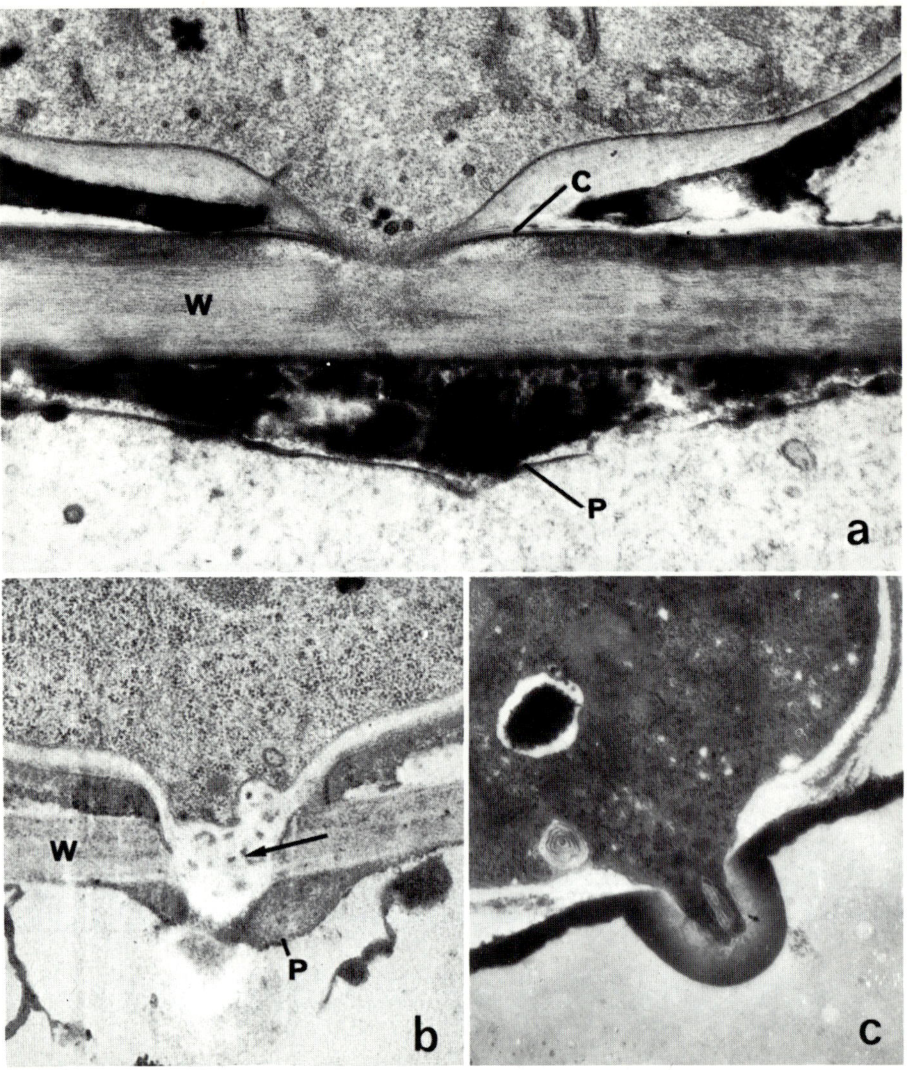

Fig. 4a—c. Appressoria and penetration pegs of *Colletotrichum* spp. (a) and (b) *C. graminicola* on susceptible maize epidermal leaf cells. Note inward depression of the cuticle *C* in (a) and cytoplasmic structures (arrow) in the cell wall of the penetration peg in (b). (c) *C. lindemuthianum* attempting to penetrate a film of cured epoxy resin. Note the sharp inward depression of the resin layer and contrast with no such effect on cell walls *W* in (a) or (b) beneath which papillae *P* have formed. (a) (×46000) and (b) (×24000). Courtesy of D. J. Politis. (c) (×18000). Reproduced by permission from Mercer et al. (1971)

Heavy metals such as uranium, which readily permeate cell walls but not protoplasts, can be used as markers to outline the protoplast surface. With this technique, fungal hyphal tips appear to be a labyrinth of intermingled protoplasmic and cell-wall components (Fig. 5a). Evidence of similar protoplasmic

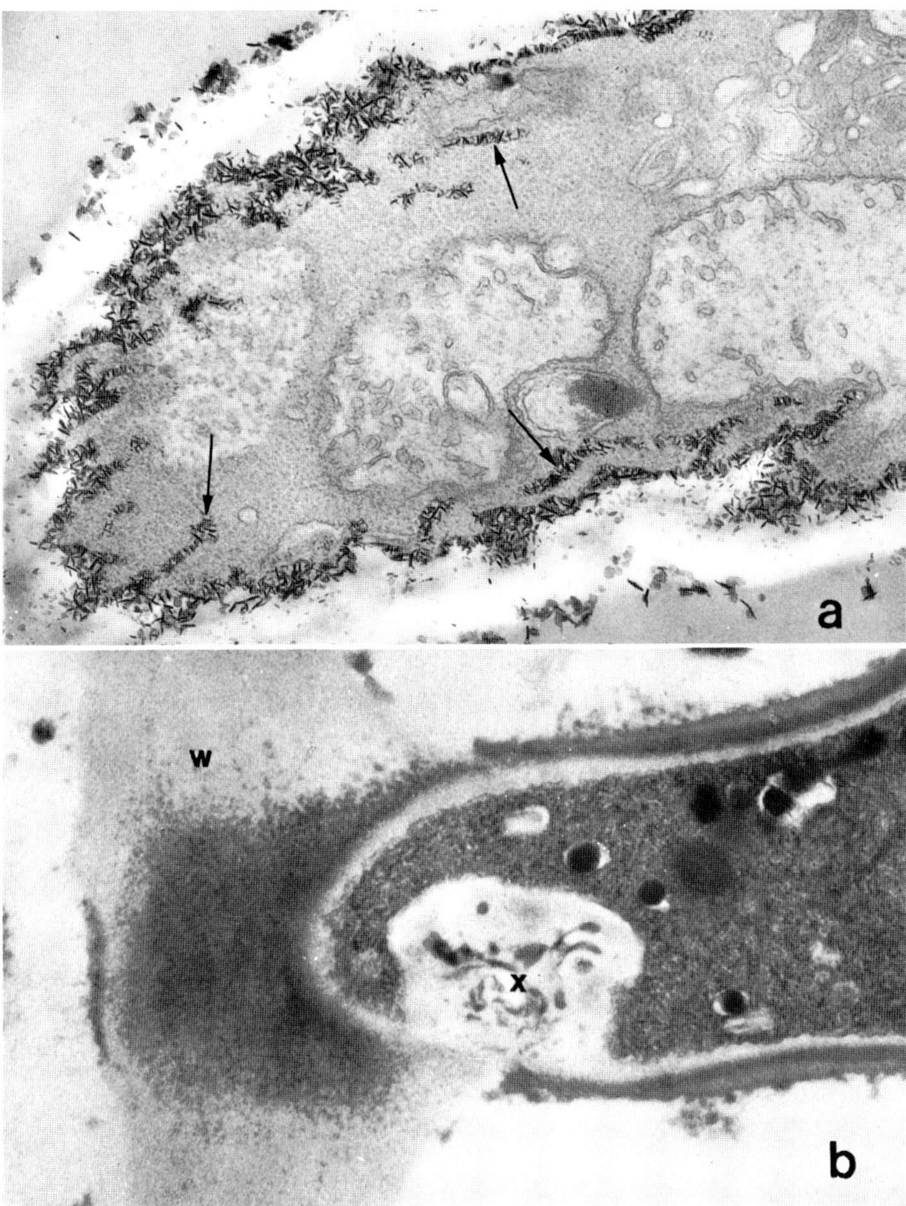

Fig. 5a and b. (a) Tip of a hypha of *Phytophthora cactorum* immersed for 1 hr in 1 mM uranyl acetate just prior to fixation. Arrows point to tentacular, crystal-bearing structures which protrude into the cytoplasm (× 52000). (b) Hypha of *Pyrenochaeta terrestris* penetrating epidermal cell wall *W* of an onion root. Note dense material in the plant cell wall and fungal wall structure (X) with tentacle-like dense material (× 33000). Reproduced by permission from Hess (1969)

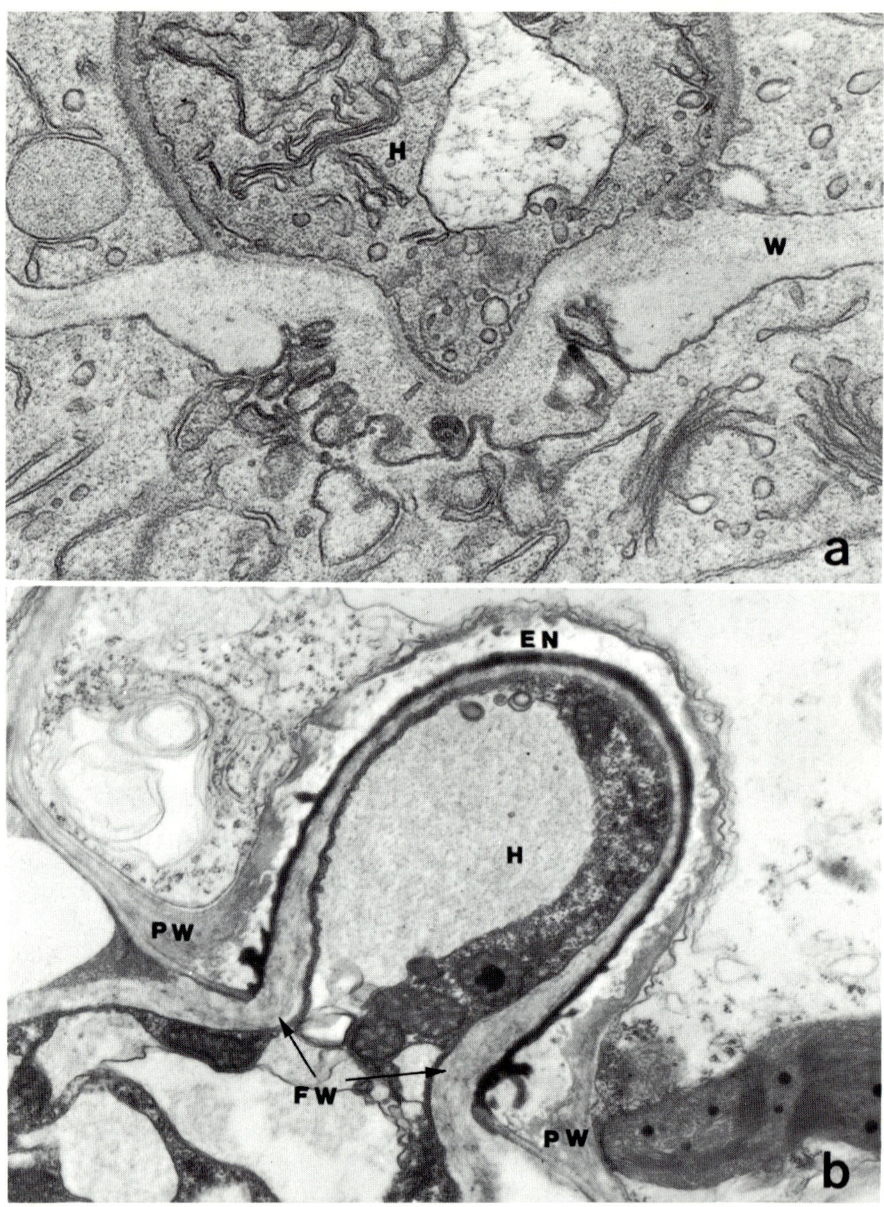

Fig. 6a and b. (a) Intracellular hypha *H* of *Phytophthora parasitica* var. *nicotianae* penetrating the wall *W* of a tobacco root cell (×42000). (b) Haustorium *H* of *Peronospora spinaciae* invaginating a spinach cell. Note that the fungal cell wall *FW* is distinct from the plant cell wall *PW* which appears softened and forced inward to form the encapsulation *EN* (×14000). (a) Courtesy of P. J. Hanchey. (b) Reproduced by permission from Kajiwara (1971)

protrusions into the cell wall can be seen in sections which graze the surface of penetration pegs and infection hyphae (Fig. 4b, 5b). Enzymes bound to finger-like or tentacular protoplasmic protrusions would appear to be ideal tools for highly localized erosion of host cell walls during penetration.

With some pathogens, enzymatic activity during cell-wall penetration appears diffuse rather than localized. Cuticular penetration by *Botrytis cinerea* is followed by rapid swelling of the epidermal cell wall (McKeen, 1974). Unlike those of *Erysiphe* and *Colletotrichum*, the tip of the penetration peg produced by *Botrytis* apparently lacks a cell wall (Fig. 3). Cell walls beneath infection cushions of *Rhizoctonia solani* also are swollen and apparently partially degraded. Such diffuse effects on cell walls may be related to the ability of these two pathogens to produce large quantities of wall-degrading enzymes.

Diffuse swelling or softening of host cell walls is also often observed during penetration by haustoria of downy mildew fungi and their allies. Haustoria produced by these pathogens appear to merely invaginate the wall without breaching this barrier during early stages of infection (Figs. 6a, b). In some cases, the host cell wall appears to have been solubilized in the area invaginated by the haustorium (Fig. 6a).

Less evidence of diffuse enzyme activity has been seen during cell-wall penetration by haustoria of rust fungi. However, rust haustoria as well as those produced by other fungi, are usually partially or completely encased by structures associated with host cell walls. These will be discussed under host responses in Chapter 3. For an extensive review of the ultrastructure of plant-pathogen interfaces see Bracker and Littlefield (1973).

An unusual type of direct penetration in which a naked protoplast is injected into the host cell has been described for members of the Olpidiaceae (Temmink and Campbell, 1969) and Plasmodiophoraceae (Keskin and Fuchs, 1969; Williams et al., 1973). Essentially the same sequence of events was observed on beet roots attacked by *Polymyxa betae* and cabbage roots by *Plasmodiophora brassicae*. Zoospores of the parasite become attached to root hairs where they encyst. The encysted spores remain apparently quiescent for about 3 hrs, during which time they form an eleborate penetration apparatus which includes a dense, bullet-shaped structure termed the Stachel. At the beginning of penetration, the cyst produces a bulb-shaped germ tube, the adhesorium, which becomes firmly adpressed to the root hair cell wall. From within the adhesorium the Stachel is forced through the hair cell wall in about 1 sec and the amoeboid parasite is then injected into the host protoplast. In view of its rapidity, Williams et al. (1973) suggest that penetration by the Stachel is primarily a mechanical process.

c) Summary of Direct Penetration

Prior to the widespread use of scanning and transmission electron microscopy, direct penetration of plant cuticles and cell walls by pathogens was thought to be primarily a mechanical process (Gäumann, 1955; Goodman et al., 1967). Although electron microscopy has provided new evidence that fungal penetration pegs are capable of exerting sufficient force to cause localized indentations in films of cured resins (Fig. 4c), there is little evidence that such pressure is required for penetration of plant cell walls. Instead, penetration pegs either appear to grow through neat holes without meeting appreciable mechanical resistance (Fig. 1 c, d; 4 b; 5 b) or they invaginate large areas where cell walls appear obviously softened or partially solubilized (Figs. 6 a, b). In both cases, enzymatic activity, either localized or diffuse, appears to play a primary role and mechanical force a secondary role in cell-wall penetration.

Evidence of the nature of cuticular penetration is more ambiguous. Cuticular membranes usually appear depressed inward during penetration (Figs. 2; 3 a), and this suggests that mechanical force has been applied. On the other hand, the neat holes produced with no signs of torn edges (Fig. 1 c, d) suggest that some softening or erosion of the rather brittle cuticle probably occurred prior to penetration. With a few exceptions (Fig. 3), present evidence suggests that cuticular penetration involves both mechanical pressure and enzymatic activity.

3. Penetration in Relation to Disease Resistance

In general, plant pathogens penetrate moderately resistant plants as readily as they do susceptible ones of the same species. Results with race T of *Helminthosporium maydis*, which destroyed millions of acres of maize during the 1970 epiphytotic of southern corn leaf blight, are typical. Maize plants which carry a cytoplasmic factor for male sterility discovered in Texas, designated Texas male sterile (Tms), are severely damaged by race T. Corresponding plants with normal (N) cytoplasm are attacked by race T but possess sufficient resistance to maintain satisfactory yields. When Tms and N plants were inoculated with spores of race T, no significant differences were found in spore germination, germ tube production, appressorial formation, or penetration of epidermal cells (Wynn, 1974). In this and in many other cases (Martin, 1964), resistance is expressed only after penetration is accomplished.

In contrast to results with race T of *H. maydis*, penetration may be completely blocked at some stage by highly resistant plants. One example, the inability of potential pathogens to penetrate even the cuticle of *Ginkgo biloba*, has already been mentioned (Chapter 1 C). Spores of *Colletotrichum piperatum* fail to produce appressoria on leaves, stems, or even green fruits of *Capsicum frutescens* which are resistant to this pathogen. On red fruits, which are susceptible, appressoria form readily (Grover, 1971). If green parts of the plant are wounded, they are readily infected. This indicates that resistance depends solely on inhibition of the formation of structures required for penetration.

Complete blockage of penetration by highly resistant tissues is probably rare. The development of haustorial pathogens on nonhosts or highly resistant hosts usually proceeds normally until or shortly after the formation of haustoria. An interesting exception is the reported inability of *Erysiphe cichoracearum*, a pathogen of cucumber, to penetrate barley, whereas *E. graminis* readily penetrates both its host, barley, and cucumber, a nonhost. Staub et al. (1974) suggest that some physical or chemical property of the wax crystals on the surface of barley leaves (Fig. 1 c) may have prevented penetration by the cucumber pathogen.

In many cases, penetration of highly resistant tissues is merely retarded rather than blocked. Maize isolates of *Colletotrichum graminicola* rapidly infect susceptible varieties of maize but produce no visible disease symptoms on oats. On susceptible maize leaves, spores germinate within 2 hrs, appressoria form within 6 hrs, and penetration of epidermal cells occurs within 9 hrs after inoculation. On resistant oat leaves, spores germination requires more than 10 hrs, appressoria form after 14–19 hrs, and cells are penetrated only after 48 hrs. Wounded oat leaves remain resistant and infection hyphae which finally successfully penetrate epidermal cells are quickly killed. This indicates that such retardation of penetration is only a first line of defense rather than a major factor in disease resistance.

In summary, evidence that barriers to penetration play a major role in disease resistance is limited to a few special cases. Generally, pathogens either penetrate resistant and susceptible plants with equal ease or penetration is merely retarded rather than blocked by resistant tissues.

B. Mechanisms of Attack

Offensive chemical weapons employed by plant pathogens include enzymes, growth regulators, and toxins. The term "employed" has been used advisedly since in many cases the source of these weapons is uncertain; both plants and pathogens are capable of producing all three types. Separation of the three classes is somewhat arbitrary since some enzymes and growth regulators produce toxic effects and some toxins possess growth-regulating properties. For our purposes, substances which exhibit enzymatic activity or well-defined growth-regulating properties will be excluded from the toxin category.

I. Enzymes

The history of research on the role of enzymes in plant pathogenesis has been traced by one of the pioneers of this area (Brown, 1965). In the past, attention has centered on enzymes which degrade plant cell walls and this trend continues. To degrade structures as complex as cell walls obviously requires a battery of enzymes acting either sequentially or in concert.

Improved methods for separating complex mixtures of enzymes have resulted in highly refined preparations capable of solubilizing individual components of plant cell walls. One of the most useful of these methods is isoelectric focusing. Fractionations by this technique are achieved by banding proteins or other amphoteric molecules in stable pH gradients maintained by electrolysis of carrier ampholytes, either in sucrose density gradients (Vesterberg and Svensson, 1966) or more often in columns of polyacrylamide gels (Finlayson and Chrambach, 1971). In such stable pH gradients, enzyme proteins migrate to their isoelectric points where they are concentrated (focused) as very sharp bands.

Classically plant cell walls have been fractionated into their major components by sequential extraction with suitable solvents. Many of the fractions so obtained are very heterogeneous and poorly defined. Highly refined enzymes capable of releasing specific wall components provide a new approach to the identification and structural relations of materials which make up cell walls. This approach, combined with gas chromatography and mass spectrometry, has provided the basis for new structural models of cell walls.

1. Structure of Plant Cell Walls

Traditionally, cellulose, hemicellulose, and pectic material in approximately equal proportions have been considered to be the major components of primary cell walls. Cellulose is a linear polymer made up of 1,4-linked β-D-glucose units. The other two classes are poorly defined mixtures of heteropolysaccharides. Fractions composed mainly of neutral hexose and pentose sugars have been considered hemicelluloses. Those made up primarily of galacturonic acids and their salts or esters have been classed pectic.

In the classical view of structure, cell walls are a mesh of crystalline cellulose microfibrils embedded in an amorphous matrix of hemicellulose with the pectic compounds localized in the middle lamella and serving to cement cells together. The first serious challenge to this view came from evidence that cell walls contain a protein rich in hydroxyproline (Lamport, 1973). Unlike the variety of proteins with enzymatic activity which had been associated with cell walls, the similarity of this hydroxyproline-rich protein to animal collagen suggested that it might play a structural role. The tenacity with which the protein resisted attempts to remove it indicated that it was crosslinked to cell-wall polysaccharides. Enzymatic degradation of the wall released a hydroxyproline-rich glycoprotein which contained arabinose and galactose. Acid hydrolysis rendered the previously protease-resistant glycoprotein susceptible to attack by trypsin. Analysis of tryptic digests led to the suggestion that tetra-arabinosides attached to hydroxyproline function to stabilize the polypeptide backbone while the major polysaccharide attachment is *via* an arabinogalactan linked to the hydroxyl group of serine (Lamport, 1973).

Polymers obtained from walls of sycamore, *Acer pseudoplatanus*, cells grown in suspension culture and treated sequentially with polygalacturonase, glucanase and pronase are listed in Table 1. Surprisingly, only seven polymeric components account for essentially the entire cell wall. Analysis of the only hemicellulose recovered, a xyloglucan, indicated that it consisted primarily of seven- and nine-sugar repeating units. The basic seven-sugar unit had four residues of β-1,4-linked glucose and three residues of terminal xylose. Each of the xylose residues was glycosidically linked to carbon six of a glucosyl residue. The nine-sugar fragment had, in addition to the basic unit, a fucosylgalactose disaccharide attached to one of the xylose residues. This xyloglucan binds firmly to cellulose fibrils apparently through hydrogen-bonding between the β-1,4-linked glucan chains of the two polymers. Only tentative structures have been proposed for the four pectic polymers recovered (Albersheim et al., 1973).

Table 1. Polymer composition of sycamore (*Acer pseudoplatanus*) cell walls. Compiled from data of Albersheim et al. (1973)

Component	% of cell wall	
Cellulose		23
Hemicellulose		
Xyloglucan		21
Pectic polymers (total)		36
Rhamnogalacturonan	16	
Arabinan	10	
4-linked galactan	8	
3,6-linked arabinogalactan	2	
Glycoprotein		19
Protein	10	
Hyroxyproline arabinosides	9	
	Total	99

The results obtained by Lamport and Albersheim and their associates have led to a cell-wall model in which cellulose fibrils are hydrogen-bonded to xyloglucans (hemicellulose) which coat their surfaces. The reducing end of the xyloglucan chains are covalently attached to pectic polymers. The latter in turn are covalently bound to the hydroxyproline-rich cell-wall protein, probably *via* arabinogalactan polymers linked to serine residues of the peptide backbone. Thus, protein, pectic, and hemicellulosic components form a network which cross-links the cellulose fibers of the cell wall.

2. Enzymatic Degradation of Cell Walls

Complete degradation of cell walls would require cellulolytic, hemicellulolytic, pectolytic, and proteolytic enzymes capable of attacking each of the major polymeric components. Although enzymes of each of these four types readily attack model substrates, only those with pectolytic activity appear effective in initiating the degradation of whole isolated cell walls. For this reason and because certain pectolytic enzymes have been associated with lethal effects during pathogenesis, this group of enzymes has received most attention.

Pectic enzymes are of two major types—pectin methylesterases which remove methoxyl groups from pectin to yield pectic acid, and chain-splitting enzymes which cleave the α-1,4-glycosidic bonds which link the uronic acid units of pectic polymers. Chain-splitting pectic enzymes are separated into two major categories; hydrolases which cleave bonds by hydrolysis and lyases or *trans*-eliminases which cleave bonds through *trans*-elimination. Members of each category are subdivided into endo- (random cleavage) and exo- (terminal cleavage) types. Further separation is based upon which substrate, pectic acid or pectin, is most readily attacked (Neukom, 1963; Bateman and Millar, 1966).

a) Role of Pectic Hydrolases and Lyases

When highly purified, enzymes capable of macerating plant tissues and initiating degradation of cell walls have been found to be polygalacturonases or pectic *trans*-eliminases usually of the endo-type. Plant pathogens are rich sources of such enzymes. The endopolygalacturonase required to initiate degradation of sycamore cell walls is the first enzyme to be secreted by the bean pathogen *Colletotrichum lindemuthianum* (Albersheim et al., 1973). Enzymes with similar activity are produced by many pathogenic fungi and bacteria, especially those that cause soft rots or water-soaked lesions (Table 2). They are also common in extracts of diseased plant tissues, but are rare in healthy tissues other than ripening fruit. These facts suggest that pathogen-produced pectic hydrolases and lyases spearhead the attack on the plant cell wall. As a result the wall structure is loosened and the remaining components are exposed to sequential attack by hemicellulases, proteases, and cellulases of plant or pathogen origin.

Crude or partially purified enzyme preparations which macerate plant tissues also kill the cells involved. Although lethal effects can be delayed by treating and holding the tissue in hypertonic solutions, extensive attempts over a period of 50 years to separate enzymatic activity from lethal effects failed (Brown, 1965). The availability of highly purified enzymes capable of rapidly macerating tissue led to renewed interest in this ancient problem. Even with the best preparations which exhibit only a single type of enzyme activity, separation of maceration from lethal effects has not been achieved. During maceration, tissues lose large amounts of electrolytes and this, like lethal effects, can be retarded by hypertonic solutions. Hall and Wood (1973) suggest that cell death caused by pectic enzymes may be an osmotic effect which results in rupture of the plasmalemma.

b) Role of Pectin Methylesterase

Unlike pectic hydrolases and lyases, pectin methylesterase (PME) is present in healthy plants bound to cell walls. Although PME production by plant pathogens has been reported, increases in PME activity in diseased tissues probably result

Table 2. Purified pectic enzymes with macerating activity associated with plant diseases

Pathogen	Enzyme[a]	Reference
Erwinia carotovora	endo-PGTE	Mount et al. (1970)
E. chrysanthemi	PGTE[b]	Garibaldi and Bateman (1971)
Sclerotinia fructigena	PMTE	Byrde and Fielding (1968)
Corticium praticola	endo-PG	Hall and Wood (1972)
Verticillium albo-atrum	endo-PG	Wang and Keen (1970)
Sclerotium rolfsii	endo-PG	Bateman (1972)
Fusarium roseum	endo-PGTE	Mullen and Bateman (1971)
Penicillium expansum	PMTE	Spalding and Abdul-Baki (1973)

[a] PGTE – polygalacturonate *trans*-eliminase; PMTE – pectin methyl *trans*-eliminase; PG – polygalacturonase.

[b] Five different fractions, three with endo- and two with exo-type activity, were obtained from two strains of the pathogen.

from increased synthesis or activation of host enzymes. On the basis of differential inhibitory effects of detergents, Drysdale and Langcake (1973) estimate that no more than 5–10% of the PME activity in *Fusarium*-infected tomato plants is of fungus origin. They conclude that it is unlikely that pathogen-produced PME has an important role in this wilt disease.

Purified preparations of PME do not cause tissue maceration or cell death. PME does not alter chain length, but demethylation brought about by its activity undoubtedly changes the properties of the cell wall and the ability of its components to resist attack by other enzymes. Bateman and Millar (1966) have suggested that demethylated pectic materials may form insoluble salts with calcium and thus become resistant to endopolygalacturonase.

c) *Role of Other Cell Wall-Degrading Enzymes*

A large number of enzymes capable of degrading isolated components of cell walls have been found in fluids on which plant pathogens have been cultured or in extracts of diseased plant tissues. These include hemicellulases (chiefly xylanase and arabinase), cellulases, cellobiases, phosphatidases, and proteases. When purified, none of these enzymes has been found to cause appreciable degradation of whole cell walls.

II. Growth Regulators

Gross symptoms alone (galls, overgrowths, epinasty, excessive elongation, witches' broom effects, etc.) implicate abnormal growth regulator activity in many plant diseases. Extracts of tissues which exhibit such abnormal growth patterns are often 10 times more active in tests for growth regulating substances than are extracts of healthy plants. Even in the absence of obvious growth abnormalities, diseased tissues often contain elevated levels of growth regulators. Furthermore, all of the major types of plant-growth regulators (auxins, gibberellins, cytokinins, and ethylene) are produced by various bacterial and fungal pathogens. These facts indicate a role for growth regulators in many diseases (see Chapter 3), but evidence that they function in initial stages of attack is very limited.

1. Auxins

Crown gall caused by *Agrobacterium tumefaciens* has served as the classic model for investigations of the role of auxins, indoleacetic acid (IAA) and related indole compounds, in plant pathogenesis. The extensive research on this disease, spurred by its similarity to carcinogenesis in animals, has been reviewed by Beardsley (1972).

Briefly outlined, tumor induction and development occur in two phases. The first, a conditioning phase, requires a wound. Conditioning is independent of temperature up to and above 32° C, and occurs in the absence of the bacterium. In the second phase, conditioned cells are transformed into tumor cells by an agent termed the tumor-inducing principle (TIP) produced by the bacterial pathogen. The transformation phase is temperature-sensitive; it takes place only below 29° C.

In tissue cultures, fully transformed tumors grow rapidly on a basal medium which contains only sugar and essential mineral elements. In contrast, tissue cultures of healthy tissues require auxin, inositol, amino acids, pyrimidines, and kinetin to maintain growth. If the transformation process is halted at various times by transfer to 32° C, tumors which grow at various rates can be obtained and their nutritional requirements compared to those of healthy and fully transformed cells. In the presence of all the organic materials required by healthy tissues, partially transformed tumors grow as fast as fully transformed ones. Moreover, slowly growing tumors do not require kinetin and more rapidly growing ones require neither kinetin nor pyrimidines. These results indicate that during transformation, cells are sequentially activated to produce substances required for autonomous growth. Growth-regulating substances produced in excessive amounts may be responsible for the uncontrolled proliferation of fully transformed tumors.

Several lines of evidence implicate auxins in tumor formation induced by *A. tumefaciens.* First, the symptoms (tumors, epinasty, lateral shoot and root development) resemble those induced by treatment with growth regulators. Second, the bacterium produces auxin in culture, and galls contain much more auxin than healthy tissues. Third, all but fully transformed tumors require an exogenous supply of auxin for continued growth in tissue culture. Finally, decapitated plants inoculated with attenuated strains of the pathogen fail to produce tumors unless auxin is added to the cut surface. Despite all this, the precise role which auxin plays in the conditioning phase or in conjunction with TIP in the transformation phase remains unknown.

Conclusive evidence of the identity of the transforming agent TIP is also lacking. At various times TIP has been suggested to be the bacterium itself, a virus associated with the bacterium, a metabolic product of the bacterium, a converted host component, or most recently, bacterial DNA. Although the hypothesis that transformation may result from uptake, incorporation and transcription of a portion of the bacterial genome by host cells is attractive it is presently supported only by circumstantial evidence. Equally uncertain is the role of undefined, cell-division promoting factors reported to be present in crown gall tissues.

The role of auxins in many other plant diseases has been evaluated by Sequeira (1973) and Van Andel and Fuchs (1972). In general, there is little evidence that pathogen-produced auxins play an important part in initiating pathogenesis or in determining the nature of plant-pathogen interactions. Instead, changes in auxin activity during pathogenesis probably result from alterations of host metabolism induced by the pathogen.

2. Gibberellins and Cytokinins

The gibberellins were discovered by Japanese workers investigating the *bakanae* disease of rice caused by *Gibberella fujikuroi (Fusarium moniliforme).* Rice plants infected by this pathogen show a striking increase in height caused by excessive elongation of internodes. A group of chemically related compounds termed gibberellins was isolated from culture filtrates of the pathogen and shown to produce symptoms typical of the *bakanae* disease. On the basis of these findings, it was reasonable to conclude that gibberellin produced *in vivo* by the pathogen was the major cause of this disease. This conclusion was weakened somewhat when gibberellins were identified as endogenous growth regulators in many healthy plants. The need for more investigations of changes in gibberellin synthesis, destruction, and interaction with other growth regulators in diseased plants has been emphasized by Van Andel and Fuchs (1972).

Evidence that pathogen-produced cytokinins may be responsible for the initiation and development of fasciation of peas caused by *Corynebacterium fascians* has been reviewed by Sequeira (1973). Extracts of cultures of the pathogen and of infected, but not healthy, tissue contain materials with cytokininlike properties. The major active component obtained from the bacterium is 6-(3-methyl-2-butenylamino) purine, the riboside of which is found in tRNA of many bacteria. Plants treated with high concentrations of kinetin develop symptoms similar to those infected by the bacterium. Although these findings suggest that cytokinins play a role in this disease, the low amount of cytokinin produced by the bacterium in culture indicates that other factors are probably involved. Further evidence, such as correlation of pathogenicity and cytokinin production among strains of the pathogen, is clearly needed to establish a pathogen-produced cytokinin as an important factor in this disease.

3. Ethylene

Ethylene appears to be the plant-growth regulator best qualified for a primary role in pathogenesis. Ethylene is ubiquitously present in plant tissues and is produced in increased amounts by damaged or diseased plants. Applied exogenously it is effective in very low concentrations (below 0.1 ppm) and produces a variety of plant responses. Some of these responses, epinasty, premature senescence, and shedding of leaves, are similar to symptoms of certain plant diseases. A number of bacterial and fungal pathogens produce ethylene and, in the case of fusarial wilt of tomato, ethylene production by the pathogen is sufficient to account for the epinastic symptoms of the disease (Dimond and Waggoner, 1953).

Despite its apparent high qualifications, ethylene as an agent in pathogenesis has been studied in only a few diseases. In the past, lack of convenient, accurate methods for measuring very small amounts of ethylene probably was responsible for much of this neglect. With rapid and highly sensitive gas chromatographic methods now available (Abeles, 1973), ethylene undoubtedly will receive more attention.

Results obtained with sweet-potato slices inoculated with the black-rot pathogen, *Ceratocystis fimbriata*, led to the hypothesis that ethylene might be a key factor which triggered metabolic changes in diseased tissues leading to resistance. Sweet-potato slices exposed to ethylene (8 ppm) for 2 days were reported to be rendered resistant to infection and, during this time, peroxidase and polyphenoloxidase increased in activity. Low concentrations of ethylene also were reported to cause marked increases in the activity of phenylalanine ammonia lyase, a key enzyme in the synthesis of aromatic antifungal and antibacterial agents. Unfortunately, further work with the sweet-potato—*Ceratocystis* model and with other systems has not confirmed these results.

The origin and possible role of ethylene in diseased plants have been discussed by Abeles (1973), Hislop et al. (1973), and Sequeira (1973). Although some bacterial and fungal pathogens produce ethylene *in vitro*, the evidence in general indicates that damaged plant tissues are responsible for increases in ethylene evolution by diseased plants. In a number of viral diseases, leaves with necrotic local lesions produced more ethylene than those from systemically infected plants without necrotic lesions. Necrosis induced by toxic chemicals also resulted in increased ethylene evolution. These results suggest that ethylene is a product, rather than the cause of, necrosis.

In culture, *Erwinia carotovora* does not produce ethylene but when cauliflower florets are inoculated with this bacterium large amounts of ethylene are evolved. Since cell-free culture filtrates of the bacterium also stimulate ethylene production in this system it is clear that the plant rather than the pathogen is the source of this gas. A pectate lyase in the culture filtrate appears responsible for increased production of ethylene by the plant (Lund, 1973).

The wide variety of physiological effects produced when ethylene is applied to healthy plants (Abeles, 1973), may account for the inconsistent and often contradictory results obtained in studies of the role of ethylene in pathogenesis. Ethylene undoubtedly interacts with other growth-regulating substances and, as Van Andel and Fuchs (1972) have pointed out, changes in the balance of growth regulators as a result of such constantly changing interactions may be more important in symptom development than the quantity of any single component.

III. Toxins

Toxins are injurious substances produced by organisms. Generally the term is restricted to substances active at physiologically low concentrations. Beyond these basic characteristics, there is little agreement on how toxins, with or without modifying prefixes, should be defined. Some follow the practice used in this chapter of excluding enzymes and growth regulators; others believe that one or both of these should be included in the toxin category. Still others restrict the term toxin to low molecular weight substances or to substances produced by microorganisms.

Opinion is equally divided on the nomenclature and classification of toxins. Systems based on the organism producing the toxin, on the organism affected by the toxin, on the toxin's chemical nature, on its mode of action, or on its effects have been proposed but none has gained wide acceptance. The term phytotoxin, meaning a toxin injurious to plants rather than one produced by plants, is in general use but some restrict the term to substances produced by pathogens whereas others include those produced by plants or plant-pathogen interactions.

Detailed discussions of terminology, definitions, and concepts of toxins in relation to plant diseases have been presented by Luke and Gracen (1972) and by Graniti (1972).

1. The Toxin Hypothesis

In its simplest form, the toxin hypothesis states that a toxic substance (X) is directly responsible for the symptoms of disease (Y). Even more briefly, X causes Y. When X is a toxin produced by a microbial pathogen, the hypothesis predicts: (a) the toxin will produce all symptoms characteristic of the disease; (b) sensitivity to the toxin will be correlated with susceptibility to the pathogen; (c) toxin production by the pathogen will be directly related to its ability to cause disease. These requirements were clearly met for the first time by a toxin produced by *Helminthosporium victoriae*, the fungus which causes Victoria blight of oats (Meehan and Murphy, 1947; Luke and Wheeler, 1955). This work, more than any other, served to establish the validity of the toxin hypothesis.

The vast majority of toxins associated with plant diseases fail to exhibit all the properties expected from the simple relationship toxin X causes disease Y. In such cases, indirect lines of evidence for a role in pathogenesis must be sought. Reproduction of particularly distinctive symptoms, demonstration of the toxin in diseased plants in quantities sufficient to cause symptoms, and the nature of the toxin and its mode of action may provide such indirect evidence.

2. The Pathotoxin Concept

Toxins which play an important causal role in disease have been termed pathotoxins (Wheeler and Luke, 1963). Pathotoxin was proposed as a broad generic term for a toxic product of a pathogen, of a plant, or of a plant-pathogen interaction. The sole requirement for pathotoxicity was convincing evidence of an important causal role in pathogenesis. Although this was clearly spelled out, the term has been widely misused by restricting it to those toxins which satisfy the simple direct hypothesis; toxin X causes disease Y. Misinterpretation probably resulted from the use of victorin, the toxin produced by *H. victoriae*, as the only clear example of a pathotoxin at the time the term was coined.

In Table 3, pathotoxins are separated into two main classes on the basis of origin. Subclasses are based on whether the pathotoxin acts selectively or nonselectively when applied to plants which are susceptible and resistant to the disease involved. The examples listed in Table 3 were selected primarily on the strength of evidence for a causal role in pathogenesis but secondarily to illustrate the system of classification. Most substances included in Table 3 are produced by pathogens. More examples of pathotoxins produced by plants or by plant-pathogen interactions could be listed if, as some prefer, enzymes and growth regulators are included in the toxin category.

The problem of providing convincing evidence of pathotoxicity was discussed by Wheeler and Luke (1963) and later by Graniti (1972), Durbin (1972) and others. Three independent, converging lines of evidence equivalent to those re-

Table 3. Classification of pathotoxins (substances which play causal roles in pathogenesis)

Class	Subclass	Produced by	Reference[a]
I. Pathogen produced	A. Selective		
	Victorin	*Helminthosporium victoriae*	Luke and Gracen (1972)
	T-toxin	*H. maydis*, race T	Hooker (1972)
	HC-toxin	*H. carbonum*	Scheffer and Yoder (1972)
	HS-toxin	*H. sacchari*	Steiner and Byther (1971)
	Phytoalternarin	*Alternaria kikuchiana*	Mori (1962)
	PC-toxin	*Periconia circinata*	Scheffer and Yoder (1972)
	PM-toxin	*Phyllosticta maydis*	Comstock et al. (1973) Yoder (1973)
	B. Nonselective		
	Fumaric acid	*Rhizopus* spp.	Mirocha (1972)
	Tentoxin	*Alternaria tenuis*	Templeton (1972)
	Marticin	*Fusarium* spp.	Kern (1972)
	Tabtoxin	*Pseudomonas tabaci*	Patil (1974)
	Syringomycin	*P. syringae*	Patil (1974)
	Phaseotoxin	*P. phaseolicola*	Patil (1974)
II. Plant or plant-pathogen produced	A. Selective		
	Amylovorin	*Erwinia amylovora* on *Rosaceae*	Goodman et al. (1974)
	B. Nonselective		
	Juglone	*Juglans nigra*	Patrick et al. (1964)

[a] References are mostly secondary sources which can be consulted for details of original works.

quired by the simple toxin X causes disease Y case (B III 1), confirmed by several different investigators, would probably satisfy even the most skeptical critics. Of the substances listed in Table 3 only victorin fully meets such stringent requirements. For the others, evidence of pathotoxicity is either less direct or it lacks independent confirmation. The pathotoxin status of the latter should be considered tentative. In general, the term pathotoxin should be used with restraint. Substances for which a causal role in disease is merely suspected rather than established can be conveniently termed phytotoxins (Wheeler and Luke, 1963). Since except for the simplest direct case, precise requirements for pathotoxicity cannot be spelled out, the final verdict on what constitutes satisfactory evidence for pathotoxicity rests on the judgment of the scientific community.

3. Selective Pathotoxins Produced by Pathogens

Pathogen-produced substances which are selectively toxic to plants susceptible to the pathogen have been termed host-specific toxins (Scheffer and Yoder, 1972). Since toxins do not have hosts, the term is a literal misnomer. More important, the word specific implies that the toxin has no effect on resistant plants or nonhosts. Actually, so-called host-specific toxins are those which at certain concentrations damage only susceptible plants but at higher concentrations also damage those that are resistant (Table 4). These toxins exhibit selective rather specific activity. They are comparable to chemicals which, properly applied, kill certain plant species without injuring others. Such chemicals are called selective herbicides. Toxins with similar properties can be properly termed selective toxins, or when appropriate, selective pathotoxins.

Victorin is by far the most potent and selective pathotoxin known (Table 4). On a dry-weight basis, refined preparations produce toxic effects on oat plants susceptible to *H. victoriae* at concentrations below 2×10^{-4} µg/ml. Moderately acid (pH 4), crude, or partly refined solutions of victorin are stable but highly refined preparations lose activity rapidly. This instability has frustrated attempts to identify victorin chemically. Present evidence suggests that victorin is polypeptide in nature and consists of several active species one of which has an isoelectric point of pH 10 (Luke and Gracen, 1972).

Victorin was discovered as the result of an outbreak of a new disease of oats in North America in the late 1940's. The pathogen, which severely damaged only oat cultivars derived from the variety Victoria, was described as a new species, *Helminthosporium victoriae*, and the disease was named Victoria blight of oats by Meehan and Murphy (1947). These workers also reported that fluids on which the pathogen had been cultured contained a toxin which caused symptoms typical of Victoria blight when applied to susceptible plants but failed to do so on those resistant to the disease. These results were confirmed and extended by Luke and Wheeler (1955) who further demonstrated that, among isolates of the pathogen, toxin production was correlated with pathogenicity. These results established victorin as a causal agent of disease and made it a valid substitute for the pathogen in studies of disease development.

Victorin, substituted for *H. victoriae*, has made it possible to follow metabolic changes in diseased plants without complications introduced by metabolic activities of a living pathogen. Since changes caused by victorin are remarkably similar to those which occur in plants infected by *H. victoriae* or many other pathogens, victorin has provided a valuable model for investigations of the nature and sequence of events which occur during pathogenesis (see Chapter 3). The earliest effect detected in victorin-treated, susceptible tissues is a marked change in permeabil-

Table 4. Potency and differential toxicity of pathogen-produced selective pathotoxins on susceptible and resistant plants

Pathotoxin	Plant	Dilution end-point on plants[a]		Selectivity
		Susceptible	Resistant	*S/R*
Victorin	Oat	1:10000000	1:25	400000
T-toxin	Maize	1:5000	1:200	25
PM-toxin	Maize	1:1000	1:25	40
HC-toxin	Maize	1:200	<1:10	> 20
Phytoalternarin	Pear	1:1500	<1:10	> 150
PC-toxin	Sorghum	1:3000	<1:10	> 300

[a] Highest dilutions of crude culture filtrates toxic to plants susceptible and resistant to the pathogens listed in Table 3.

ity. This led to the suggestion that disruption of cell permeability may be an initial event which triggers subsequent pathological changes in many diseased plants (Wheeler and Luke, 1963).

Other than victorin, the pathotoxin which has received most attention is T-toxin produced by *Helminthosporium maydis* race T. Although a pathotoxin produced by *H. maydis* had been reported earlier (Smedegard-Peterson and Nelson, 1969), the disastrous outbreak of southern corn leaf blight in 1970 made this pathogen a subject of intense interest. Texas male sterile (Tms) plants, which are highly susceptible to race T (Chapter 2AII3), are 25 times more sensitive to T-toxin than normal resistant plants (Table 4). Male sterility *per se* caused by the cytoplasmic Tms factor does not account for disease reactions since Tms plants in which male fertility has been genetically restored remain susceptible to the disease and sensitive to T-toxin. Several substances, some of which are thought to be terpenoids, have been reported to exhibit selective toxicity when injected into leaves of Tms and normal plants (Strobel, 1974). Unlike crude or partially refined culture filtrates, these do not exhibit selective toxicity in root-growth tests which have been used extensively to separate Tms from normal plants. Further work is clearly necessary to establish the chemical nature of the active principle in T-toxin.

Crude or partially refined solutions of T-toxin cause rapid swelling and loss of respiratory control when added in low concentration to mitochondria isolated from Tms plants. Mitochondria from resistant normal plants are not affected even when much higher concentrations of T-toxin are used (Miller and Koeppe, 1971). These reults suggested that the initial effect of T-toxin might be exerted directly on the respiratory centers. Further work, however, failed to provide evidence of an early effect of T-toxin on mitochondria *in situ* when intact tissues were treated. Instead the earliest effect detected, a change in the electrochemical potential of treated cells, suggests that like victorin, the initial effect of T-toxin is on cell permeability.

Preliminary results indicate that PM-toxin, produced by *Phyllosticta maydis* which causes yellow leaf blight of maize, acts in essentially the same way as *H. maydis* T-toxin. Plants with Tms cytoplasm are more susceptible to *P. maydis* and to PM-toxin than those with normal cytoplasm. Selective effects of PM-toxin on mitochondria and on permeability similar to those caused by T-toxin have also been reported (Comstock et al., 1973; Yoder, 1973). Whether the two toxins are similar chemically remains to be determined.

Like victorin, selective pathotoxins produced by *Helminthosporium carbonum* (HC-toxin) and *Periconia circinata* (PC-toxin) appear to be polypeptide in nature. The HC-toxin may be a cyclic molecule containing unsaturated amino acid residues. There is evidence that crude preparations of PC-toxin contain two or more selectively toxic agents with different chemical properties (Pringle, 1972). Otherwise these toxins resemble poor carbon copies of victorin. They are much less potent and selective (Table 4) and their ability to induce physiological changes similar to those in diseased plants is less well documented. The progeny of crosses of *H. victoriae* and *H. carbonum* have been reported to segregate 1:1:1:1 for pathogenicity to oats, to corn, to both, and to neither, and in the same way for production of victorin, HC-toxin, both, and neither (Scheffer and Yoder, 1972). Thus a single gene pair in each pathogen controls both pathogenicity and toxin production.

Preliminary results indicate that a selective toxin is produced by *Helminthosporium sacchari* which causes eye-spot disease of sugarcane (Strobel, 1974). Culture filtrates of *H. sacchari* produce reddish-brown stripes, called runners, when injected into susceptible leaves. Tests of 182 clones of sugarcane indicated a statistically significant correlation between disease susceptibility and sensitivity to the toxin. However, disease and toxin ratings were not closely correlated for 33 of these clones. A toxic fraction named helminthosporoside (proposed structure 2-hydroxycyclopropyl-α-galactopyranoside) has been reported to bind to a single protein present in susceptible but not in resistant clones. Resistant clones, however, contain a similar protein which binds helminthosporoside after the protein is treated with a detergent.

If these results can be independently confirmed and extended, helminthosporoside should become a valuable model for investigations of pathogenesis and the nature of disease resistance. Evidence that this toxin plays a causal role in pathogenesis would be greatly strengthened if toxin production could be correlated with pathogenicity and if the toxin could be shown to produce more of the symptoms observed in diseased plants. The runners, which are the only symptom reproduced by the toxin, are less distinctive and occur after the formation of eye spots from which the disease gets it name. Furthermore, if this toxin is to serve as a model system, evidence that it produces physiological changes similar to those found in diseased plants should be provided.

Culture filtrates of *Alternaria kikuchiana* are selectively toxic to Japanese pears *(Pyrus serotina)* susceptible to the black-spot disease caused by this fungus. Drops of culture filtrates applied to leaves cause black spots typical of the disease on susceptible but not on resistant plants. Three selective toxins designated phytoalternarins A, B, and C, were obtained from culture filtrates but have not been identified chemically. Two other species, *A. citri* and *A. mali* have been reported to produce selective toxins (Otani et al., 1974).

4. Nonselective Pathotoxins Produced by Pathogens

Three main lines of evidence have been used to implicate nonselective toxins in plant disease. These are reproduction by the toxin of distinctive early disease symptoms, correlation of toxin production and pathogenicity, and recovery of the toxin from diseased plants in quantities sufficient to account for symptom development. Nonselective pathotoxins are those for which at least two of these three lines of evidence have been provided. Such evidence of pathotoxicity is obviously less conclusive than that provided for selectively toxic agents.

Species of *Rhizopus* associated with the hull-rot disease of almonds produce large amounts of fumaric acid when grown on almond hulls or various other substrates. On diseased plants, leaves near rotted hulls are blighted and adjoining twigs are killed although they are not invaded by the fungus. These leaf and twig symptoms can be reproduced by applying fumaric acid to fruit mesocarps, and before these symptoms develop fumaric acid accumulates in high concentrations in leaves. After symptoms develop, only small amounts of fumaric acid are found in diseased leaves. Mirocha (1972) suggests that these results can be explained on the basis of the epoxysuccinate cycle. Fumaric acid, synthesized in the infected mesocarp, is translocated to leaves and twigs where it is first metabolized to *trans*-epoxysuccinate which is toxic and then further metabolized to *meso*-tartaric acid. The latter is quickly converted to oxalacetate which enters the tricarboxylic acid cycle and other metabolic pathways. This hypothesis, which accounts for the initial build-up and later disappearance of fumaric acid, is supported by evidence that epoxysuccinate applied to mesocarps is even more effective than fumarate in inducing symptoms on twigs and leaves.

Tentoxin is produced by *Alternaria tenuis*, and both the toxin and the pathogen cause striking variegated chlorosis in seedlings of cucumber, cotton, citrus and other plants. Tentoxin has been reported to cause large reductions in chlorophyll content, to inhbit cyclic photophosphorylation, and to cause stomatal closure in plants sensitive to the toxin. Cabbage plants which are insensitive to the toxin do not show these effects. The latest structure proposed for tentoxin is cyclo-N-methyl-dehydrophenylalanyl-L-N-methylalanyl. The mechanism responsible for tentoxin-induced chlorosis remains uncertain since some investigators failed to obtain reductions in chlorophyll content in toxin-treated sensitive plants. Work on tentoxin has been reviewed in detail by the leading investigator of this toxin (Templeton, 1972) and more briefly by Strobel (1974).

Marticin is one of several red pigments with a naphthazarin structure produced by *Fusarium* species of the *Martiella* group. Highly pathogenic strains of the pea pathogen, *F. solani* f. sp. *pisi* produce large quantities of marticin in culture whereas only traces of this compound are produced by weakly pathogenic strains. Furthermore, marticin has been extracted from diseased plants in quantities sufficient to cause wilting and general necrosis (Kern, 1973). Although this indicates that marticin plays a role in disease, the evidence would be strengthened if the symptoms produced were more distinctive.

Evidence for causal roles in pathogenesis for toxins produced by bacterial pathogens has been summarized by Patil (1974). For those listed as pathotoxins in Table 3, the case rests on ability of the toxin to reproduce distinctive disease symptoms and correlation of toxin production with pathogenicity. For syringomycin, where toxin production may not always be correlated with pathogenicity, the case is strengthened by evidence that a substance similar to syringomycin can be recovered from diseased plants.

Tabtoxin, formerly known as the wild-fire toxin, is produced by *Pseudomonas tabaci* which causes wild-fire of tobacco. For many years this toxin produced by *P. tabaci* was thought to be a natural antimetabolite of methionine because its toxicity to *Chlorella* could be overcome by methionine. The chief problem with this hypothesis was the failure of methionine to protect tobacco plants against the effects of the toxin or the pathogen. Work with partially refined preparations indicated that the toxin inhibited glutamine synthetase activity and that tobacco leaves infiltrated with glutamine were protected from the effects of the toxin. This led to the hypothesis that toxin-induced chlorosis resulted from ammonia accumulation caused by inhibition of glutamine synthetase which catalyzes one pathway of nitrogen assimilation. Although ammonia accumulates in toxin-treated leaves, later work with highly refined toxin and enzyme preparations gave no evidence that the toxin inhibits glutamine synthetase activity. Thus the mode of action of this toxin remains in doubt. Tabtoxin has been identified chemically as β-lactam-threonine. This toxin or a closely related compound, β-lactam-serine, is also produced by at least two other plant pathogenic species of *Pseudomonas* (Patil, 1974).

Less information is available for phaseotoxin produced by *P. phaseolicola*, a pathogen of beans, or for syringomycin produced by *P. syringae* which attacks stone-fruit trees. Both toxins are thought to be peptides. Patil (1974) has suggested that phaseotoxin-induced chlorosis may be the result of citrulline deficieny caused by inhibition of ornithine carbamoyl-transferase since leaves pretreated with citrulline do not produce chlorotic halos when toxin is injected. Syringomycin is a potent, broad spectrum antibiotic which may act by disrupting cellular permeability (Backman and DeVay, 1971).

5. Pathotoxins Produced by Plants or by Plant-Pathogen Interactions

Fire-blight of apples and other species of *Rosaceae* is a highly destructive disease caused by *Erwinia amylovora*. Goodman et al. (1974) have reported that the slime or "ooze" exuded from slices of green apples inoculated with virulent strains of the pathogen is selectively toxic to plants susceptible to the disease. This selective toxin is not produced by virulent or avirulent bacteria grown on a defined liquid medium, and it is not recovered from the small amount of ooze produced when apple slices are inoculated with avirulent strains of *E. amylovora*. A refined preparation of the toxin contained 98% galactose in a polymeric form and ca. 0.4% protein. The molecular weight was calculated to be about 165000. The toxin, considered to be a product of an interaction between the plant and the pathogen, has been named amylovorin (Table 3). Although amylovorin has been included in the pathotoxin category, its selective toxicity is based solely on wilting effects which cannot be considered a distinctive symptom of disease. Pending confirmation and further evidence the pathotoxin status of amylovorin remains tentative.

The leaves, bark, and green fruits of black walnut *(Juglans nigra)* trees contain large quantities of a nonphytotoxic compound, hydrojuglone. This compound is readily oxidized to juglone (5-hydroxy-1,4-naphthoquinone) upon exposure to air. Juglone is highly toxic to a number of plants and is believed to be responsible for failure of such plants to grow in the vicinity of walnuts. Other examples of phytotoxic materials produced by plants or by microbial breakdown of plant products are discussed by Patrick et al. (1964). These have not been included since this general area is beyond the scope of this book.

6. Phytotoxins

A very large number of substances, herein designated phytotoxins, have been implicated to some degree as causal agents of plant diseases. Many phytotoxins have been discovered only recently and some of these undoubtedly will be reclassified as pathotoxins when more conclusive evidence for a causal role in pathogenesis is made available. The role of other phytotoxins in disease remains uncertain despite extensive investigations. A few examples of phytotoxins will be discussed briefly. More extensive discussions can be found in books and reviews (Wood et al., 1972; Luke and Gracen, 1972; Templeton; 1972).

A number of terpenoid phytotoxins have been described. One of these, a glucoside of a carbotricyclic terpene named fusicoccin, is produced by *Fusicoccum amygdali* which attacks peach and almond trees. This toxin possesses growth-regulator properties and causes stomates to open. This effect on stomates may be responsible for wilting symptoms of the disease but additional evidence is required before pathotoxicity can be claimed.

Fusaric acid is the classic example of a pathogen-produced toxin which meets the criteria proposed for vivotoxins as defined by Dimond (1955). Fusaric acid is produced by several species of *Fusarium* which cause wilt diseases. Its chemical structure (5-n-butylpicolinic acid) is well established and it accumulates in toxic quantities in diseased plants. Introduced into healthy plants it causes wilting and necrosis (Kern, 1972). However, despite extensive investigations over a 20-year period, the significance of fusaric acid as a factor in pathogenesis has not been established.

Chapter 3 Responses of Plants to Pathogens

Although changes in structure, function, and metabolism which occur during pathogenesis will be discussed in separate sections, they are obviously interrelated and inseparable in diseased plants. Any change in one may cause or be caused by a change in one or both of the others. Perhaps more important than the changes themselves is the sequence in which they occur. In this chapter, special attention will be given to the few diseases in which the sequence of all three types of changes have been studied with particular emphasis on those that occur during early stages of pathogenesis.

A. Pathological Alterations in Structure

I. The Sequence of Changes in Infected Living Tissues

The classical pathogen *Phytophthora infestans*, which causes late blight of potatoes, has been used by Tomiyama (1971) and his associates in an extensive series of light microscopic studies of responses of living cells to infection. Potato cultivars which carry R genes are hypersensitively resistant when inoculated with incompatible races of *P. infestans* but are susceptible to compatible races. When cut surfaces of petioles or tubers are inoculated with zoospores, penetration begins within 15 min and the contents of the infection structures move into the host cells during the next 20 min. These events occur in essentially the same way in both compatible (susceptible) and incompatible (resistant) combinations and also on surfaces inoculated immediately after cutting and on those aged in moist chambers for 5 hrs or more before inoculation. However, when aged tissues are inoculated with an incompatible race, cells are killed very rapidly; 30% die within 20 min and 60% within 40 min after penetration. In contrast, when freshly cut surfaces are inoculated with an incompatible race, cell death is greatly delayed; only about half of the penetrated cells are dead 8 hrs after inoculation. In both aged and freshly cut tissues, cells penetrated by a compatible race remain alive for at least 2 days while the fungus ramifies through the tissues.

Rapid cell death following inoculation of aged tissues involves a define sequence of events. In initial stages of penetration a striking increase in protoplasmic strands occurs and the nucleus of the cell moves to the site of penetration. This is followed by the appearance of cytoplasmic particles in rapid Brownian movement. At this time the cells are judged to be still alive since they stain with neutral red and can be plasmolyzed. Cell death is signaled by a granulation of the cytoplasm and the appearance of many more particles in Brownian movement. Later the cell contents become yellow and finally, after Brownian movements cease, dark brown. For several hours after cell death occurs, hyphae of the pathogen grow as rapidly in hypersensitively resistant tissues as in susceptible ones in which cells are not killed.

Cell death following penetration by incompatible races of *P. infestans* can also be greatly delayed by treating the tissues with metabolic inhibitors such as sodium azide. Tomiyama (1971) suggests that a similar metabolic paralysis caused by cutting may account for the delay in cell death observed when freshly cut tissues are inoculated. On this hypothesis, hypersensitive reactions would be governed by the level of host metabolic activity.

Infection of grass root hairs by zoospores of *Pythium aphanidermatum* also results in very rapid cell death (Kraft et al., 1967). Shortly after a germ tube of the zoospore makes contact with the root hair, an appressorium forms and within 20 min produces a penetration peg. As penetration proceeds, cyclosis in the hair cell slows and then ceases about 35 min after penetration. Five minutes later the cytoplasm becomes granulated and 20 min later it undergoes false plasmolysis (see Fig. 12).

Pathological responses of living cells have also been studied by interference microscopy combined with histochemical procedures. In onions inoculated with *Botrytis allii*, cells adjacent to fungal hyphae showed decreases of 28% in nuclear area and of 40% in nuclear dry mass. Cells 2 cm away showed slight increases in both of these characters. Results with onions inoculated with *Aspergillus niger* were similar except that larger increases in nuclear area and nuclear dry mass were found 2 cm away from hyphae of the pathogen (Pappelis and Kulfinski, 1971). These results obtained with weak parasites differ from those observed in rust-infected wheat. Nuclei in rust-affected cells increased in size and this was accompanied by a two-fold increase in RNA (Whitney et al., 1962). In all cases studied, cell death was accompanied by pycnotic nuclear collapse and large losses of DNA.

II. Pathological Changes in Ultrastructure

Although transmission electron microscopes can reveal structural details impossible to resolve by light microscopy, interpretation of electron micrographs presents several problems. Electron microscope images of cells which have been killed, embedded, sectioned, and stained with heavy metals obviously are artifacts which must be interpreted with caution. Since continuous or sequential observations on the same cells are impossible, the sequence of changes during infection must be deduced from less direct evidence. Usually samples are taken at various times after inoculation but changes are rarely sufficiently synchronous to allow determination of the sequence of events by this method alone. Another approach is based on the rationale that if one change (X) is often observed in the absence of a second change (Y) but Y is never observed in the absence of X, then X preceded Y.

1. Changes Associated with Plant Cell Walls

a) Papillae

The earliest and most consistent morphological response to pathogens which penetrate directly (Chapter 2II2b) is the formation of structures called papillae.

Papillae usually are localized in the area directly beneath the penetration peg (Figs. 2, 4a,b, 7, 8) but they may also form in adjacent uninvaded cells if penetration occurs at or near cell junctions (Fig. 9). Papillae vary from small, dense, amorphous deposits (Figs. 2, 4) to large complex structures which contain membranous inclusions (Figs. 7, 8). Large papillae, which can be seen with light microscopes, have been subjected to histochemical tests. Results of such tests have been highly variable. Some indicate that callose is the main component, others that lignin, suberin or other materials predominate. The only consistent result has been a positive test for polysaccharides.

The time at which papillae form, whether before, during, or after cell walls have been penetrated, has received much attention. The clearest case appears to be infection of cabbage root hair cells by *Plasmodiophora brassicae* where the process has been followed by light microscopy of living tissues. Here the papilla is not present until after the Stachel (Chapter 2AII) has penetrated the cell wall. It then forms very rapidly within two minutes after cell-wall penetration. On the other hand, papilla-like structures form on epidermal walls of cells of *Ginkgo biloba* in response to fungi incapable of penetrating even the cuticle (Chapter 1C). Evidence from electron micrographs indicates that papillae are present before cell walls arc completely penetrated but not before infection pegs are formed (Politis and Wheeler, 1973). Sections through appressoria or other infection structures which miss the small area of penetration may give the false impression that papillae are formed prior to the formation of the infection peg. For example, if the section shown in Fig. 7a had missed the tip of the penetration peg, a massive papilla would have appeared to have formed prior to penetration.

Several lines of evidence suggest that mechanical forces applied by the pathogen during penetration may provide the stimulus for papilla formation. First, papillae are restricted to areas beneath or immediately adjacent to the point of penetration. Second, papilla-like structures can be induced by gentle pin-pricks. Third, when no evidence of localized mechanical force is found during penetration, papillae are not formed (Figs. 3, 10). Finally, papilla-like structures are not formed in response to treatment with victorin which causes other morphological changes typical of disease.

Papillae may impede or block penetration by some pathogens but evidence for such a role is far from conclusive. The fact that infection hyphae do not enlarge greatly until papillae have been penetrated suggests that these structures offer some resistance to lateral expansion of the pathogen (Figs. 7a, 9). In some cases, penetration pegs produce lateral branches after elongation of the peg in its initial direction has apparently been halted within the papilla (Fig. 8). On the other hand, infection pegs often penetrate quite massive papillae and the occurrence and size of these structures does not appear to be correlated with disease resistance.

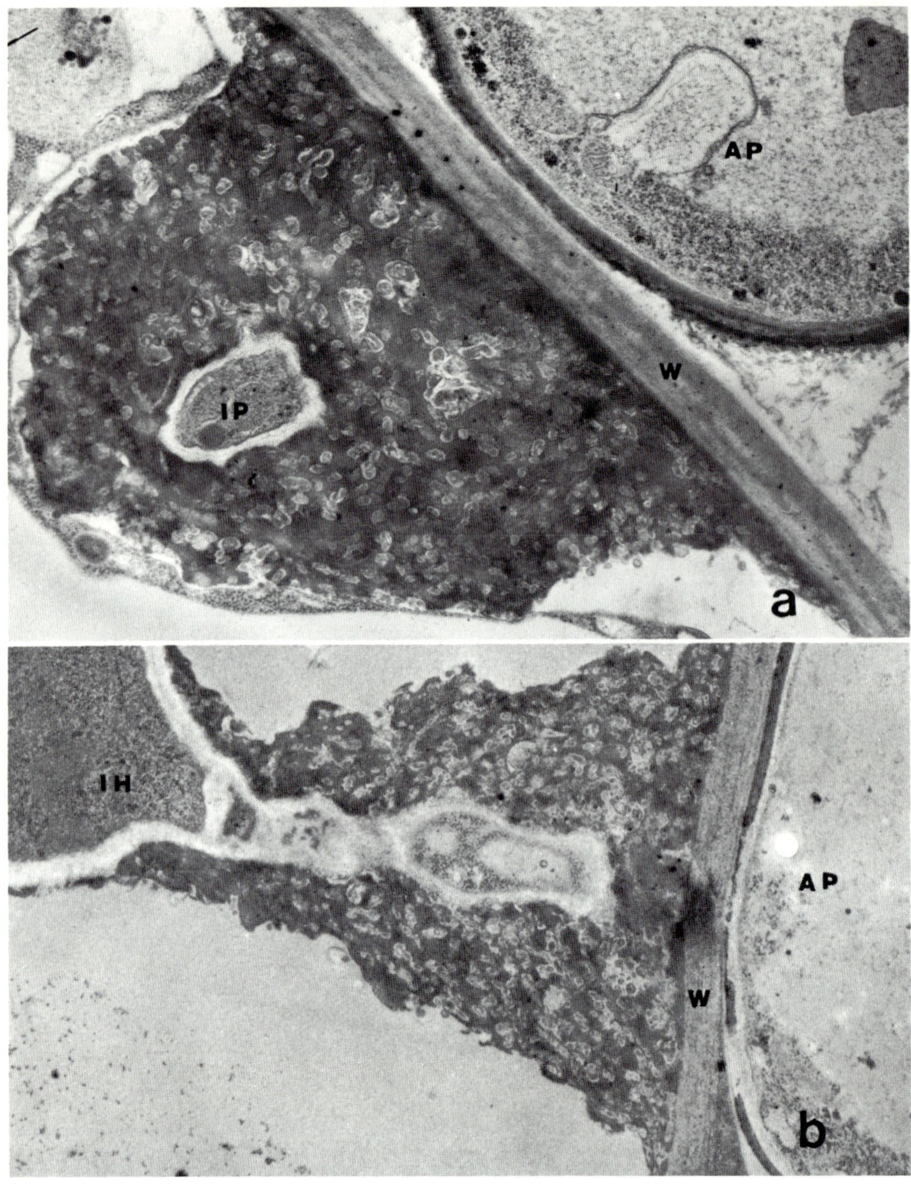

Fig. 7a and b. Massive papillae formed on the inner surface of epidermal cell walls *W* of maize leaves 9 hrs after inoculation with *Colletotrichum graminicola*. Only portions of the appressoria *AP* are shown in these nonmedian sections. In (a), the tip of the infection peg *IP* remains embedded in the papilla, whereas in (b) the papilla has been penetrated and an infection hypha *IH* formed (both ×24000). Courtesy of D. J. Politis

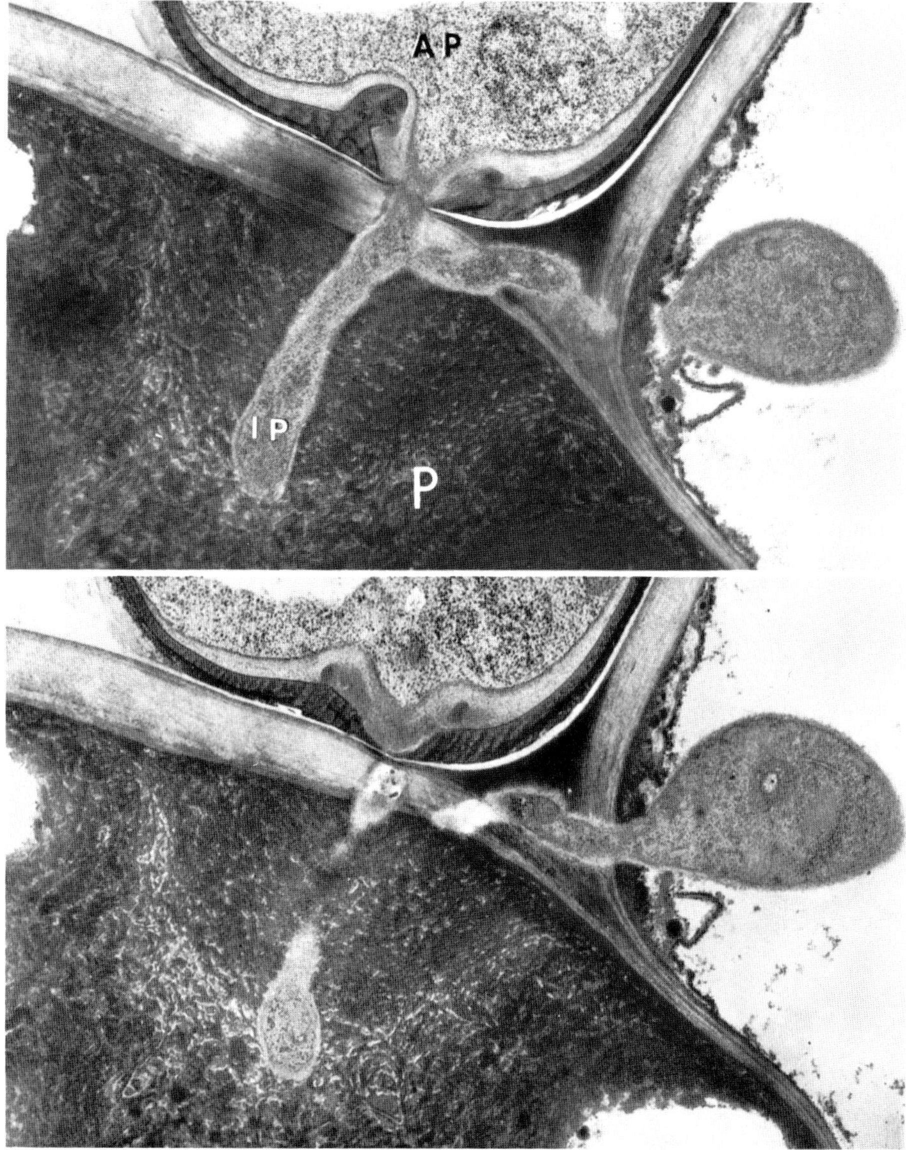

Fig. 8. Two sections through infection structures produced on highly resistant oat leaves 72 hrs. after inoculation with *Colletotrichum graminicola*. Growth of the initial penetration peg *IP* appears to have been halted within the papilla *P* and a branch has formed which has penetrated an adjacent cell. Note that the inner wall of the appressorium *AP* appears buckled inward as if in response to backthrust pressure (× 11000). Courtesy of D. J. Politis

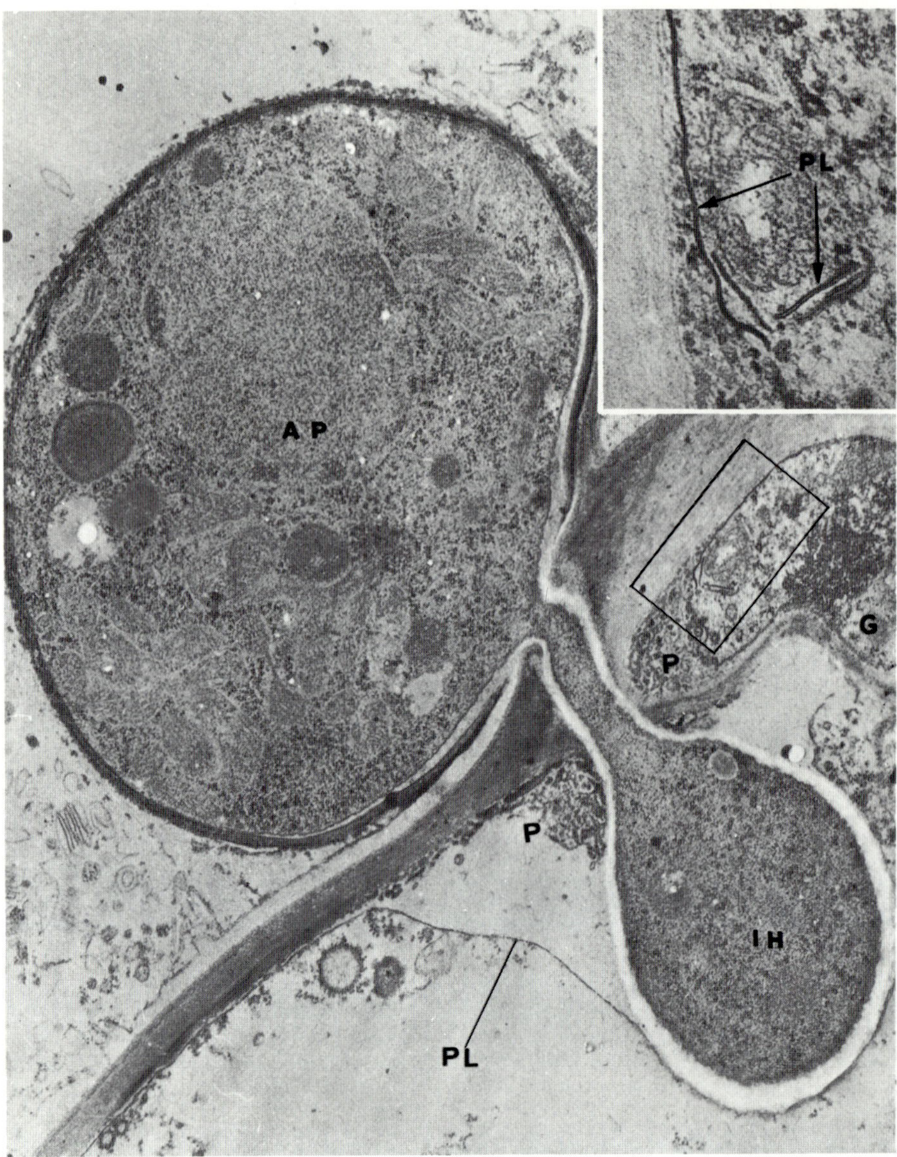

Fig. 9. Penetration of an epidermal cell of a maize leaf by *Colletotrichum graminicola* 9 hrs after inoculation. The infection hypha *IH* which has invaginated the plasmalemma *PL* of the host cell is surrounded by an electron-lucent wall which is continuous with the inner wall of the appressorium *AP*. Papillae *P* are present both in the invaded cell and the adjacent, noninvaded guard *G* cell (× 22000). Inset, enlarged from boxed area showing disrupted plasmalemma (× 81000). Reproduced from Politis and Wheeler (1973)

It was pointed out in Chapter 2AII3 that in some cases barriers to penetration appear to retard the infection process in resistant plants. This is well illustrated by a comparison of infection structures produced by *Colletotrichum graminicola* on highly resistant oat leaves (Fig. 8) with those formed on susceptible maize (Fig. 9). The most obvious difference is the time required for penetration, less than 9 hrs on susceptible maize but more than 48 hrs on resistant oats. A second difference is the thickness and conformation of the electron-lucent inner wall that forms in the floor of the appressorium. On resistant oats it is much thicker than on maize and shows inward bucklings that suggest that much more mechanical force has been applied. Third, on maize this inner wall extends to form a sheath around the infection hypha, whereas on oats such extension either does not occur or, if it does, the sheath is so thin as to be barely detectable. Finally, branched penetration pegs (Fig. 8) have never been observed in infections of susceptible maize.

The nature and possible function of the distinctive electron-lucent wall which surrounds infection hyphae in susceptible cells (Fig. 9) has been discussed by Politis and Wheeler (1973). They point out that it serves to preclude any direct physical contact between the protoplasts of the plant and the pathogen, and in this respect, at least, it resembles the encapsulations which surround haustoria of biotrophic pathogens. The fact that it is greatly reduced or absent in resistant interactions (Fig. 8) suggests it may function in maintaining a compatible relationship during early stages of infection of susceptible tissues.

b) Cell-Wall Modifications

In plants invaded by fungi, viruses, or nematodes, cell walls are often swollen, distorted or otherwise modified. These modifications range from small, blister-like lesions which arise just beneath the inner surface of the wall (Fig. 10) to very large elaborately contoured structures which encase haustoria (Fig. 6b). Considering the great variation in size and morphology it is not surprising that many different terms have been used for these structures. Structures not directly associated with haustoria have been called wall-lesions, blisters, lomasomes or lomasome-like, paramural bodies, or boundary formations. Those surrounding haustoria have been termed sheaths, encapsulations or encasements. All of these structures have certain features in common. They are either continuous with or a part of host cell walls and they consist of an electron-lucent matrix in which dense amorphous or membranous materials are embedded. Since the function and nature of these structures are not known, Bracker and Littlefield (1973) have suggested that the term cell-wall appositions be used for structures other than those surrounding haustoria. Even this term suggests that these structures are built up by secretion of cytoplasmic materials and, in many cases, no evidence of such secretion has been produced. Instead, swelling or partial solubilization of the cell wall appears to be responsible for at least initial changes in cell-wall structure. For this reason, an even more noncommittal term, cell-wall modifications, will be used for these structures associated with host cell walls.

Swollen or otherwise modified cell walls are often associated with initial stages of penetration by fungal pathogens (Figs. 3, 6a, 10a). Unlike papillae they are not

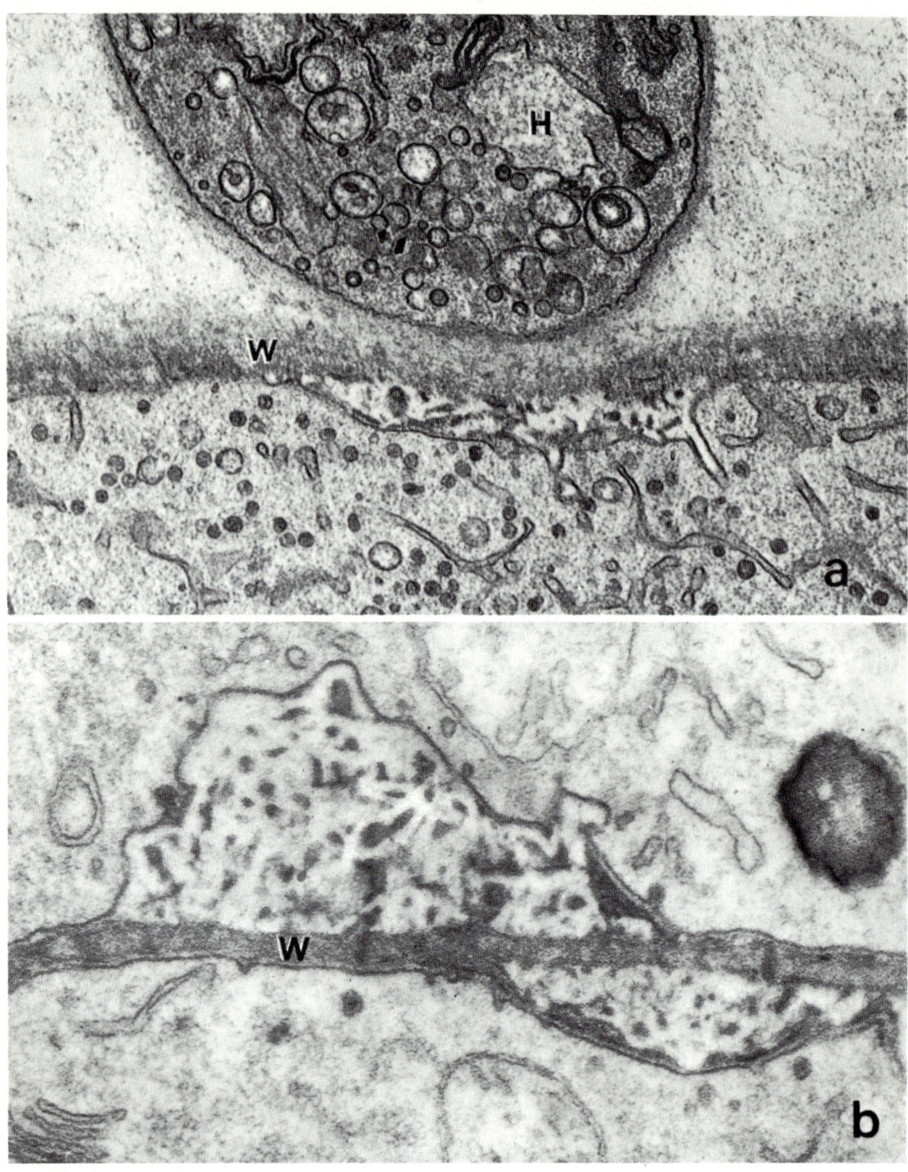

Fig. 10a and b. Cell-wall modifications in material fixed in potassium permanganate. (a) Small structure formed in the wall *W* of an epidermal root cell of tobacco during penetration by a hypha *H* of *Phytophthora parasitica* var. *nicotianae* (× 58000). (b) Larger modification in the wall of an oat root cell induced by treatment with victorin, the pathotoxin produced by *Helminthosporium victoriae* (× 55000). (a) Courtesy of Penelope J. Hanchey. (b) Reproduced from Hanchey et al. (1968)

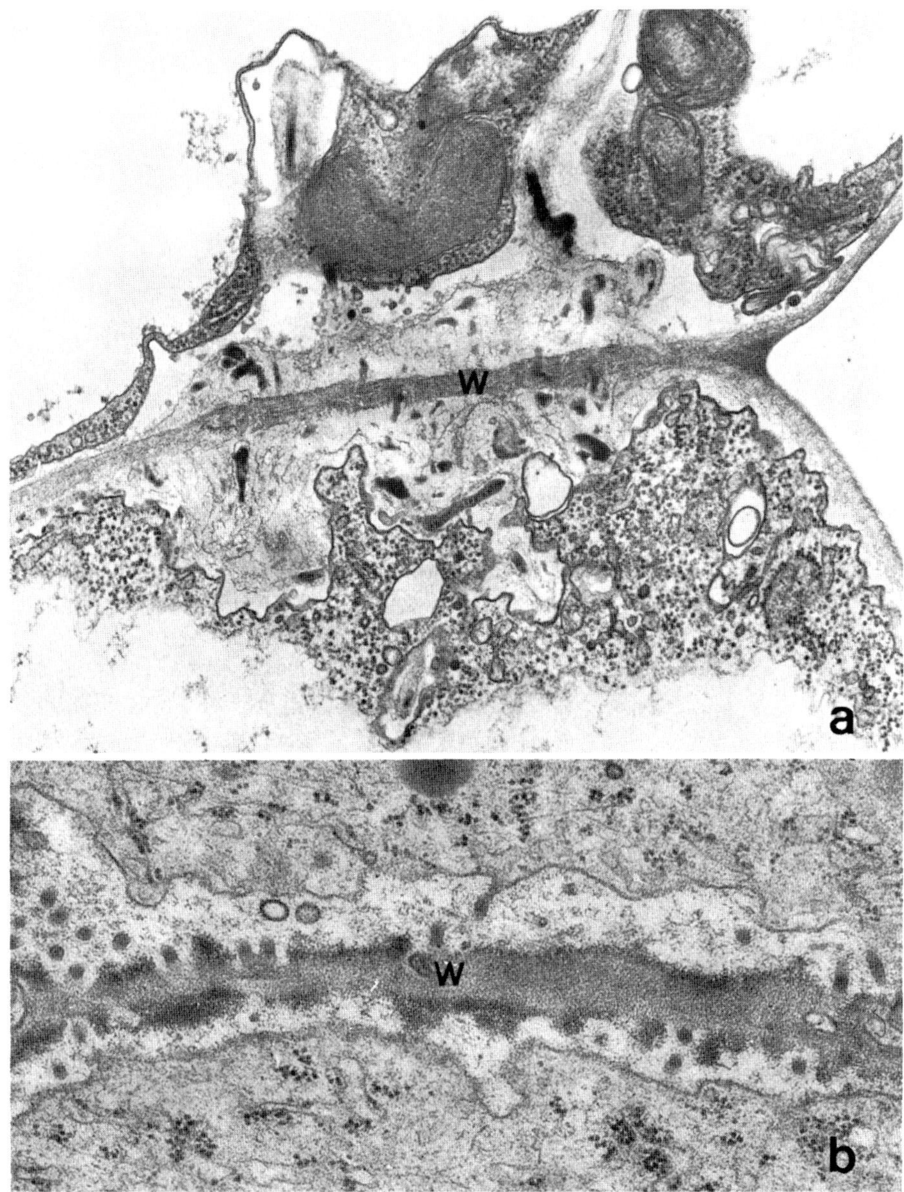

Fig. 11 a and b. Cell-wall modifications in material fixed in glutaraldehyde followed by osmic acid. (a) Modified cell wall *W* in a bean leaf systematically infected with cowpea mosaic virus (× 30000). (b) Modified wall *W* believed to represent callose deposition near a necrotic local lesion induced by potato virus X on a leaf of *Gomphrena globosa* (× 28500). (a) Courtesy of J. P. Fulton and K. S. Kim. (b) Reproduced from Allison and Shalla (1974)

restricted to the area of penetration and their formation does not depend on mechanical forces since they are found in tissues treated with pathotoxins (Fig. 10b) or infected with viruses (Fig. 11). Thickened, swollen or otherwise distorted cell walls are very common in areas immediately surrounding local necrotic lesions caused by viruses. Histochemical tests have indicated the presence of callose in these walls and this has led to suggestions that plasmodesmatal connections may have been sealed off and the spread of the virus prevented. Although this may account for failure of virus particles to spread in local lesions host, it should be noted that highly modified cell walls have also been observed in plants systemically infected with viruses (Fig. 11a).

Cell-wall modifications similar to those found in infected plants can be induced in roots exposed to victorin (Hanchey et al., 1968) or to moderately toxic concentrations of uranyl salts (Wheeler, 1974). In both cases, cells with highly modified walls but otherwise healthy in appearance are often found completely surrounded by necrotic or heavily damaged cells with few or no cell-wall modifications. This led these workers to suggest that cell-wall modifications might provide localized protection from the effects of toxic agents. A similar role has been postulated for structures which encase haustoria (Ehrlich and Ehrlich, 1971).

2. Pathological Changes in Membranes and Cellular Organelles

Increased secretory activity appears to be consistently associated with various types of cell-wall modifications observed in diseased plants. In many cases, the secretory vesicles have been shown to be products of the Golgi apparatus. Golgi secretory products are a major component of extracellular slime excreted by roots and may play an important role in cell-wall synthesis (Morré and Mollenhauer, 1973). It is therefore likely that Golgi secretion products contribute to the swollen and distorted cell walls found in diseased plants. In oat roots treated with victorin, cell-wall lesions occurred before increases in Golgi activity were observed, but Golgi products apparently played a role in the development of these lesions into large characteristic cell-wall modifications. The possible role of secretory vesicles in the formation of papillae and other structures associated with plant-pathogen interfaces and in exchanges of materials between haustoria and the cells they invade has been discussed by Bracker and Littlefield (1973).

Disruption of cell permeability which is often the first change detected in diseased tissues (3. B. I), suggests that ultrastructural changes in membranes, especially the plasmalemma, may occur during initial stages of pathogenesis. In a few cases, discontinuities in the plasmalemma have been observed in cells during early stages of infection (Fig. 9, inset), but in most cases such discontinuities or other evidence of gross disruption have been found only in cells that were obviously heavily damaged and in advanced stages of degeneration. The only early effect on the plasmalemma of oat root cells treated with victorin was a change that made unit structure of the membrane easier to resolve (Hanchey et al., 1968). This suggests that rapid losses of electrolytes caused by victorin result either from very subtle changes in membrane structure or from effects on sorptional or other properties of the cell which are not related to the membrane surface.

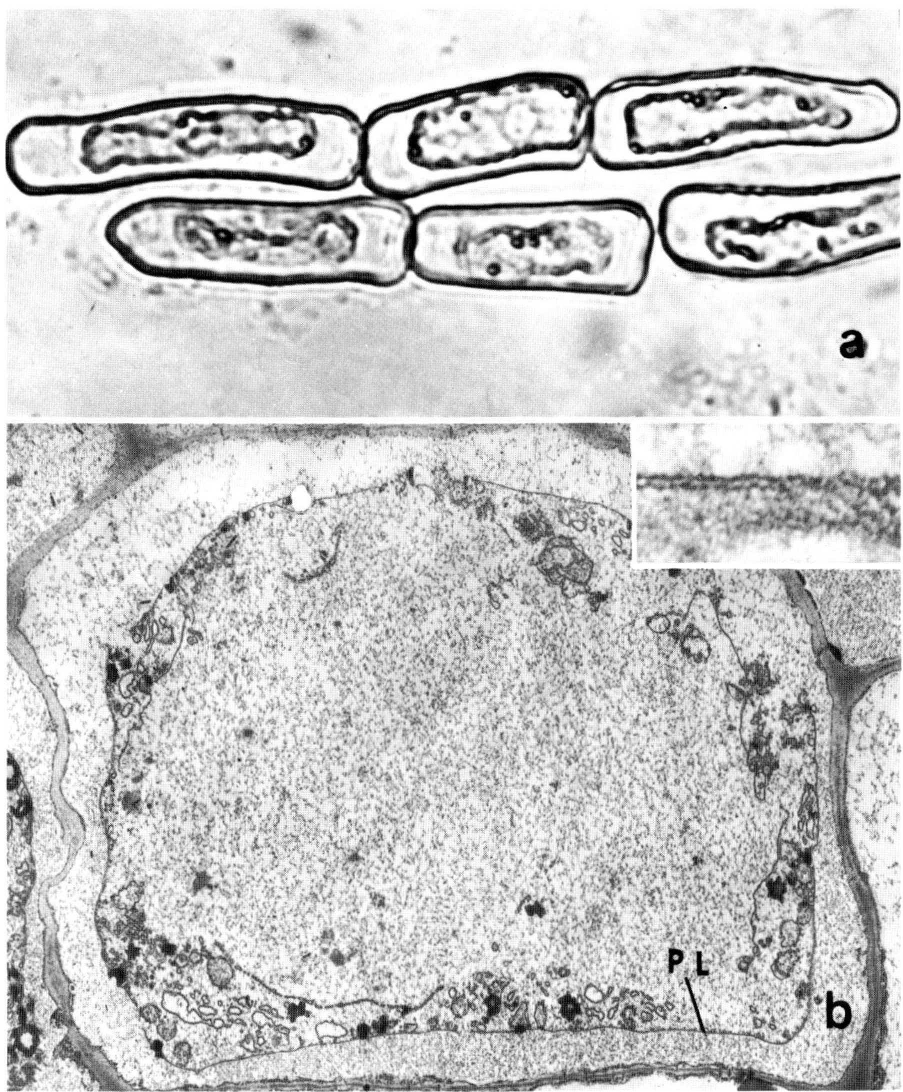

Fig. 12a and b. False plasmolysis in root cap cells of oats treated with victorin. (a) A group of loose nonfixed cells (× 300). (b) An interior root cap cell. Note that the plasmalemma *PL*, though discontinuous, is relatively intact and that its unit membrane structure can been seen in the inset (× 9000). Inset (× 250000). Reproduced from Hanchey and Wheeler (1969)

Other early changes in ultrastructure are observed only in certain types of diseases. Where chlorosis is an early symptom, disruption of chloroplast structure is often the first change detected. In tissues invaded by biotrophic fungi and by certain nematodes, large increases in ribosomes have been reported (Webster, 1969; Bracker and Littlefield, 1973).

3. False Plasmolysis

In infected or otherwise damaged tissues, cells often suddenly plasmolyze. Since the phenomenon occurs when cells are held in hypotonic solutions or even in distilled water it has been called false plasmolysis. One example, false plasmolysis as the final death stage in cells attacked by zoospores of *Pythium*, has already been cited (3.A.I). In general, false plasmolysis occurs in a highly erratic and unpredictable fashion. However, when susceptible oat roots are treated with victorin, false plasmolysis occurs regularly in virtually all epidermal and outer root cap cells (Fig. 12). Results of a light and electron microscope study of victorin-treated oat roots (Hanchey and Wheeler, 1969) indicated that all membranes in cells which had undergone false plasmolysis were more or less disrupted. Surprisingly, the plasmalemma appeared least affected (Fig. 12b). These workers suggested that disruption of the tonoplast and mixing of vacuolar and cytoplasmic contents would result in rapid loss of osmotic properties and cell death. Shrinkage of the protoplast (false plasmolysis) would then be brought about by weak attractive forces which are thought to cause membranes to coalesce after rupture.

B. Pathological Alterations in Function

I. Changes in Cell Permeability

Regardless of disease type or the nature of the pathogenic agent, changes in permeability have been found in all diseased tissues which have been examined. Furthermore, in time-course studies, changes in permeability have been detected before any other symptom of disease appeared. This, plus the fact that changes in permeability can be caused by a variety of biological, chemical, or physical agents, led to the suggestion that a change in permeability might be the initial event in pathogenesis (Wheeler and Luke, 1963).

Permeability changes in diseased plants can be followed by one or more of several methods. Changes in osmotic potentials which accompany changes in permeability can be determined by placing tissues in solutions of varying osmotic potential. Solutions of sucrose or mannitol are often used and the concentration required to produce plasmolysis in 50 per cent of the cells is considered to be isotonic. The method can be varied by first plasmolyzing the cells in solutions 3 times isotonic and then measuring the time required for deplasmolysis either in the same solutions or in solutions of lower osmotic potential. With this method vacuolated cells are required and if vacuolar contents are not pigmented, neutral red or some other dye may be added to aid detection of changes in protoplast volume. The method allows comparisons of individual cells of the plant and the pathogen and of plant cells in healthy aud diseased areas. To obtain satisfactory quantitative data, large mumbers of cells must be examined and the method becomes quite tedious.

Changes in permeability can be followed by measuring substances released by tissues bathed in distilled water or other solutions. Amounts of materials released can be determined by chemical analysis of samples of the bathing solution, or, in the case of electrolytes, by measuring electrical conductivity of the bathing solution with a conductivity bridge. The conductivity method has been popular because of its simplicity, but results should be interpreted with caution since materials capable of chelation may be released which mask the electrolytes. Measurements of radioactivity released from tissues previously labeled with radioactive substances is an alternative, and highly sensitive, method for dermining changes in permeability.

Changes in electrochemical potentials, measured by inserting the tip of a microelectrode into a cell with a reference electrode held in the external bathing solution, have been used to detect permeability changes caused by pathotoxins. With this method, changes in potentials, usually called membrane potentials, can be detected within seconds. Since the tip of the internal electrode is usually inserted into the vacuole, the measured current must traverse the tonoplast, a layer of groundplasm, the plasmalemma, and the cell wall. Under these conditions, the term membrane potential, if used, should not be interpreted as specifying any particular component of the system.

1. Nature of Permeability Changes in Diseased Plants

Recent investigations of the role of permeability changes in pathogenesis have been carried out mostly with enzymes or pathotoxins. This work will be discussed in later sections. Here only a brief summary of earlier investigations, reviewed in detail by Wheeler and Hanchey (1968), will be presented.

Permeability changes characteristic of diseased plant tissues occur as early, if not initial, responses to a variety of pathogenic agents. In initial stages of attack cells become leaky. They release large quantities of electrolytes and other substances and at the same time lose the ability to accumulate mineral salts. Loss of water from such cells probably accounts for the water-soaked condition which is often the first visible symptom of disease. The leaky condition persists even after respiratory rates have increased sharply. This indicates that the normal relationship between respiration and uptake of materials by cells has been disrupted.

In susceptible reactions characterized by diffuse spreading symptoms, affected tissues including those not invaded by the pathogen are usually found to be leaky in late as well as early stages of disease. In contrast, cells which border necrotic local lesions on resistant plants have been found to be less permeable than healthy cells and accumulation of various substances by these cells indicates they have an increased rather than a decreased uptake ability. These results provided the basis for the hypothesis that permeability changes may govern disease reactions by regulation of the availability of substances required for growth and development of the pathogen. It should be noted that this hypothesis does not account for reactions to highly selective pathotoxins such as victorin. Instead, resistance to victorin appears to depend on low sensitivity to the toxin and the ability to recover from its initial effects (Luke and Gracen, 1972).

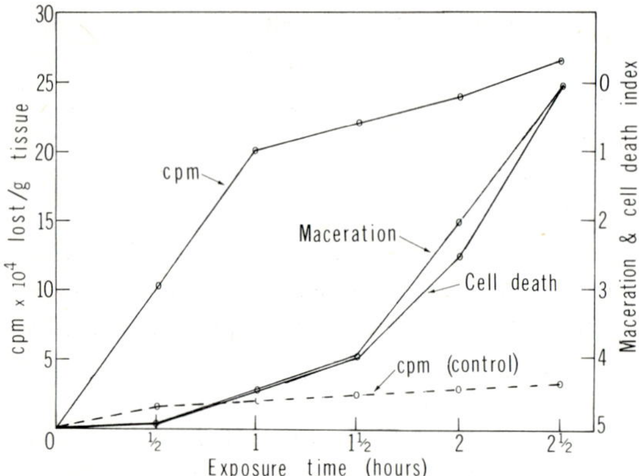

Fig. 13. Leakage of ^{86}Rb (cpm), maceration and cell death in potato-tuber disks treated with an endopolygalacturonate *trans*-eliminase from *Erwinia carotovora*. Redrawn from Mount et al. (1970)

2. Changes in Permeability Induced by Pectic Enzymes

All purified pectic lyases and hydrolases capable of causing cell maceration (Table 2) also cause cell death. In time-course studies, cell death and maceration are closely correlated but both are preceded by marked changes in permeability. Potato disks treated with an endopolygalacturonate *trans*-eliminase produced by *Erwinia carotovora* provide a typical example of these relationships (Fig. 13). Cell death and maceration were not detectable after 30 min. of exposure to the enzyme but both were detected after 60 min and the two paralleled each other for the remainder of the experiment. In contrast, drastic changes in permeability, evidenced by large losses of ^{86}Rb from the tissue, occurred within 30 min. Isotope loss continued at a rapid rate for an additional 30 min and then more slowly as cell death and maceration increased (Fig. 13). These results show that permeability changes not only precede cell death but also that most of the isotope is released by cells judged to be alive on the basis of their ability to retain neutral red.

Results very similar to those in Fig. 13 have been obtained with crude extracts of potato tubers rotted by *E. carotovora* or by *Corticium praticola*. Extracts of *C. praticola*-rotted tissues were particularly effective. They caused large increases in losses of electrolytes within 10 min with no effects on tissue coherence until after 50 min (Hall and Wood, 1973). Maceration and cell death were also closely correlated when potato disks were treated with a purified pectin methyl *trans*-eliminase produced by *Penicillium expansum* (Spalding and Abdul-Baki, 1973).

Although tissues exposed to pectic enzymes undergo changes in permeability before cells are separated or killed, the relationships among these phenomena are far from clear. Cell death and permeability changes can be greatly delayed if tissues are plasmolyzed in hypertonic solutions during exposure to pectic enzymes but such conditions have little effect on the maceration process (Brown,

1965). Thus maceration does not appear directly related to either altered permeability or to cell death even though it is correlated with the latter in time.

Further evidence that cell death may be only indirectly related to maceration has come from recent work with isolated protoplasts. Earlier studies had indicated that protoplasts, stabilized in hypertonic media, were killed when exposed to a crude enzyme preparation with macerating activity but less rapidly than intact cells of tissue slices. This appeared consistent with delayed cell death in tissue slices plasmolyzed in hypertonic solutions. Protoplasts were also killed but more slowly by heat-inactivated enzyme preparations. These results suggested that both a heat-labile substance, presumably a pectic enzyme, and a heat-stable one might be involved (Brown, 1965).

Protoplasts have now been exposed to highly purified preparations of endo-PGTE, phosphatidase C, and protease (Tseng and Mount, 1974). Of these, only endo-PGTE caused rapid cell death and tissue maceration of both potato and cucumber slices. After long exposures (3 hrs or more) the other two enzymes killed a few cells of cucumber without maceration but had no effect on potato slices. When isolated cucumber protoplasts stabilized in 0.6 M sucrose were exposed to these enzymes, endo-PGTE had no effect but the other two caused protoplasts to collapse. Phosphatidase C, which killed 50% of the protoplasts in 1 hr, was approximately twice as effective as protease. Although these results implicate enzymes other than pectolytic in cell death they must be reconciled with evidence that proteases and phosphatidases have no significant effects on the viability of protoplasts obtained from other species of plants. Furthermore, since work with protoplasts requires plasmolyzing conditions, the possibility remains that endo-PGTE or other pectic enzymes may be directly lethal to unplasmolyzed cells. Finally, in contrast to results with potato slices (Fig. 13), cucumber tissue labeled with ^{86}Rb did not release significant quantities of this isotope when treated with endo-PGTE for a period of 4 hrs. This is particularly strange in view of the fact that all cells were judged dead by the neutral red test after 1 hr of exposure to the enzyme.

3. Changes in Permeability Induced by Toxins

The role of toxin-induced permeability changes in pathogenesis has been discussed in several reviews (Luke and Gracen, 1972; Page, 1972; Scheffer and Yoder, 1972; Wheeler and Hanchey, 1968). Since by definition, toxins are injurious substances which damage and kill cells, there is no question about their ability to disrupt permeability. The questions which need to be considered are: do toxins disrupt permeability directly or indirectly, and are these disruptions the cause or the result of other pathological events during pathogenesis? The most extensive search for answers to these questions has been carried out with victorin, the pathotoxic product of *Helminthosporium victoriae* (Table 3).

Details of the discovery of the striking effect of victorin on permeability and of later studies of the nature of this effect can be found in the references cited in the preceding paragraph. In brief, victorin causes changes in permeability which are entirely similar to those characteristic of diseased plant tissues (see the preceding

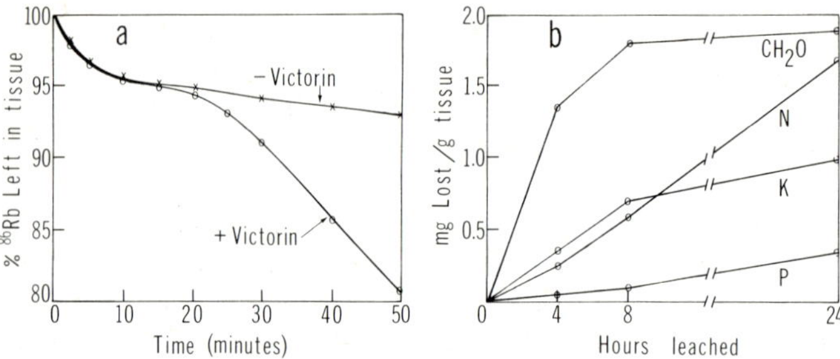

Fig. 14. (a) Loss of ^{86}Rb from oat roots exposed at time zero to victorin. (b) Loss of carbohydrates (CH_2O), nitrogenous materials *N*, potassium *K*, and phosphorus *P* from oat leaves pretreated with victorin and then shaken in water. (a) Redrawn from Keck and Hodges (1973); (b) from Black and Wheeler (1966)

section). Susceptible tissues exposed to low concentrations of victorin become leaky and lose the ability to accumulate salts and other materials. Similar effects can be obtained with resistant tissues but only with much higher concentrations of victorin (Table 4).

Initial results showed that losses of electrolytes from victorin-treated tissues were virtually instantaneous whereas increases in respiration and other changes occurred later after a lag of about 30 min. Applied directly to isolated mitochondria, victorin had no effect on permeability, oxygen uptake, or respiratory control ratios. Victorin also failed to cause swelling or other changes in chloroplast preparations. Analysis of materials leached from victorin-treated tissues shaken in distilled water showed that potassium was the chief mineral element lost (Fig. 14b). Since the bulk of intracellular potassium is thought to be localized in vacuoles, losses of large quantities of this element indicated disruption of tonoplast permeability.

Taken together, these initial results suggested that pathological increases in respiration and subsequent events might be secondary effects brought about indirectly by disruption of tonoplast permeability. Potassium or other salts released from the vacuole into the cytoplasm could result in salt stimulated respiration similar to that induced when salts are supplied exogenously. Release of phenolics or other toxic materials from vacuoles might be responsible for eventual cell death. Further evidence that victorin disrupts tonoplast permeability has come from compartmental analysis of ion efflux from roots labeled with ^{86}Rb by Keck and Hodges (1973). They concluded that victorin-induced losses of this isotope from the cytoplasm and from the vacuole occurred almost simultaneously about 20 min after treatment (Fig. 14a).

In an attempt to test the hypothesis that victorin might act by disrupting tonoplast permeability, victorin-treated tissues were leached in distilled water. These experiments were based on the rationale that if potassium or other materials leaking from the vacuole were responsible for victorin-induced increases in respiration, leaching might remove the respiratory stimulators from the cells and inhibit the respiratory increase. Initial results appeared to support the hypothesis.

The duration of elevated respiratory activity induced by victorin was reduced from 24 hrs to less than 8 hrs by leaching. However, addition of potassium or other salts to the solutions in which tissues were bathed failed to overcome the leaching effect. Although those results do not rule out the hypothesis, a more critical test is obviously needed.

Further attempts to determine the nature of victorin-induced changes in permeability have been designed to detect initial events. The earliest effects detected by electron microscopy at the cell wall-plasmalemma interface (3.A.II.2), led to the suggestion that victorin exerted its initial effect at the protoplast surface. Some support for this suggestion came from evidence for selective toxic effects of victorin on isolated protoplasts (Samaddar and Scheffer, 1968). However, unlike losses of electrolytes from tissues, effects of victorin on protoplasts were not instantaneous. When victorin was applied to susceptible protoplasts at concentrations which completely inhibited root growth, 50% of the protoplasts remained intact for 1 hr. With concentrations 1000 times higher, effects on protoplasmic streaming could be detected in 10 min and protoplast collapse somewhat later. Thus effects on protoplasts appear to occur with a lag similar to, but shorter in duration, than that required for changes in respiration.

Victorin-induced losses of ^{86}Rb also occurred only after a lag of about 20 min (Fig. 14a). Losses of isotope during the first 20 min, which were not affected by victorin, were attributed to wash-out from cell wall free space. Since rubidium and its analog potassium are thought to behave similarly in efflux experiments, these results indicate that victorin-induced losses of potassium would also require an induction period of about 20 min. These results must be reconciled with evidence of immediate effects of victorin on permeability which have been confirmed by short-term measurements of electrochemical potentials (Fig. 15c) and of losses of electrolytes from coleoptiles (Fig. 15b).

The initial pattern of victorin-induced electrolyte loss (Fig. 15b) is remarkably similar to that of victorin-independent loss of ^{86}Rb (Fig. 14a). In both cases, efflux is rapid during the first 5 min. It then declines sharply, fluctuating in the case of electrolytes, until a new steady rapid rate of loss becomes etablished with victorin-treated tissues. This pattern, and especially the decline in rate after the first 5 min, would not be expected if the initial effect of victorin were a disruption of protoplast permeability. Instead the initial effect appears to involve cell walls. After a lag of about 5 min, victorin causes a distinct but transient increase in elongation of susceptible oat coleoptiles (Fig. 15a). This effect, which is much more rapid than response to auxins, clearly signals a change in cell walls.

Considered together the data from short-term experiments indicate an initial effect of victorin on some component of cell walls followed after 20–30 min by a second effect on protoplasts. Whether measured by changes in electrochemical potential (Fig. 15c) or by losses of electrolytes (Fig. 15b), the initial phase is essentially completed within 5 min. Losses of rubidium, which should reflect losses of potassium, are not affected by victorin during this initial phase (Fig. 14a). Hence, large losses of potassium in longer-term experiments (Fig. 14b) probably result from disruption of protoplast permeability, which occurs only after a lag of about 20 min.

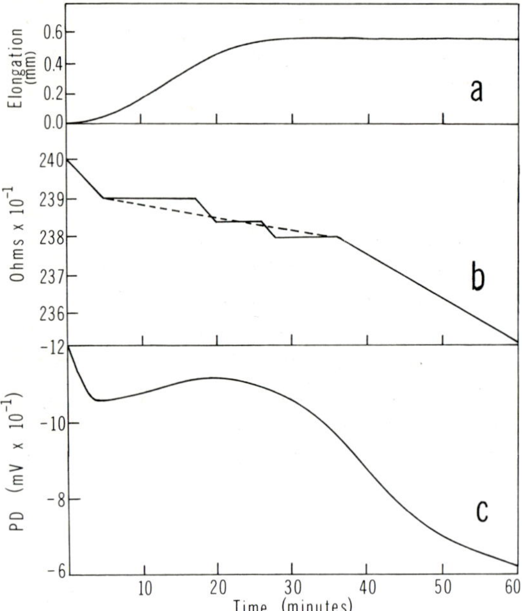

Fig. 15a—c. Short-term responses to victorin applied at time zero to sensitive oat tissues. (a) Transient increase in elongation of coleoptiles. (b) Loss of electrolytes from coleoptiles estimated from changes in electrical resistance of a bathing solution. The dashed line is an average rate during a period of fluctuation. (c) Changes in electrochemical potential of a syncytium induced in an oat root by the nematode *Heterodera avenae*. Such syncytia are well suited for monitoring with inserted microelectrodes since they are much larger than normal cells which respond to victorin in the same way. (a) and (b) Redrawn from Saftner and Evans (1974); (c) from unpublished data supplied by A. Novacky and M. D. K. Jones

In general, results of short-term experiments do not support the hypothesis that initial changes in permeability induced by victorin are caused by disruption of the plasmalemma. Instead, protoplast permeability assumed by membrane theory to be controlled by the plasmalemma, is disrupted only after an induction period more than ample to allow the toxin to enter and act within the protoplast. Although such a lag would be expected on the basis of sorptional theories of permeability, such theories have little support in the Western Hemisphere. Also consistent with sorption theory but difficult to explain on membrane theory is the pattern of loss of potassium (Fig. 14b) and rubidium (Keck and Hodges, 1973) in long-term tests. In both cases, losses occurred at essentially a constant rate for a period of 8 hrs. Victorin-treated tissues lost ^{86}Rb after a 20-min lag as if the cells had only one compartment. If, as membrane theory assumes, these mineral elements are mostly free in two compartments, the cytoplasm and the vacuole, sharp sequential losses would be expected as victorin disrupted first the plasmalemma and then the tonoplast.

In general, all available data indicate that the initial effect of victorin is on permeability but the site and nature of this effect remain uncertain. Clarification of the mechanism of action of this remarkable substance should shed new light not only on physiological aspects of pathogenesis but also on the more general problem of the nature of cell permeability.

II. Impairment of Water Relations

Imbalances in water relations will obviously be one result of alterations in cell permeability which were discussed in the preceding section as characteristic of diseased plants in general. Localized water soaking is often the first visible symptom of disease, and in later stages diseased tissues often dry out rapidly and appear withered even in those diseases not classified as wilts. However, most work on water relations has been done with diseases in which wilting is the major symptom. At least three major causes of pathological wilting can be distinguished; inability to absorb water, interference with water transport, and stomatal dysfunction.

1. Interference with Water Absorption

Most varieties of pepper plants infected with tobacco etch virus (TEV) develop vein clearing and leaf mottling but do not wilt or die. Tabasco pepper *(Capsicum frutescens)* plants infected with TEV develop the usual leaf symptoms but these are followed in a few days by a sudden wilt and eventual death. A histological study of TEV-infected Tabasco pepper plants at the onset of wilt revealed a ring of necrotic phloem and cambium cells around the xylem of the roots. No such necrosis was found in stems, petioles or leaves, and no clogged xylem vessels could be found in any part of the plant (White and Horn, 1965).

Resistance to water movement in the xylem as an important factor in TEV-induced wilting was ruled out by the fact that wilted plants regained turgor when their root tips were excised under water (Ghabrial and Pirone, 1967). These workers also found that increased losses of electrolytes, especially potassium, occurred 24–48 hrs before wilting or phloem necrosis could be detected. They suggested that these permeability changes might initiate further reactions which sealed off a ring of injured necrotic cells and prevented water from reaching the xylem. In contrast to these results, wilting caused by certain root-rotting fungi can be reversed only when stems are excised above necrotic areas (Duniway, 1973). In such cases, abnormally high resistance to water movement in the xylem may be responsible for wilting.

2. Interference with Water Transport

Fungi and bacteria that cause vascular wilts invade and become established in the water-conducting xylem vessels of the plant. In general, fungal wilt pathogens remain confined in the xylem until the plant is moribund, whereas bacteria often escape to form cavities or pockets in adjacent tissues. Within the xylem, bacterial cells or fungal propagules are carried by the transpiration stream into petioles and leaves where they germinate to produce fresh infections. Such transport accounts for the rapid spread of the pathogen and the development of systemic infections even in tall trees. Resistance to *Fusarium* wilt of banana has been associated with the rapidity with which the plant produces gels and tyloses that prevent migration of the pathogen (Beckman, 1964). Such a host response does not account for resistance to the Dutch elm disease or to a variety of other wilt diseases (Dimond, 1970).

The mechanism responsible for the symtoms of vascular wilt diseases has been and continues to be the subject of vigorous debate. Two general mechanisms have been proposed: failure of the vascular system to transport an adequate supply of water, and disruption of leaf-cell permeability by wilting toxins. Those who support vascular dysfunction as the major cause of wilting cite evidence that resistance to water flow in diseased plants is greatly increased. In segments of stems infected by various wilt pathogens, resistance to water flow is commonly increased 4 to 60 times and in *Fusarium*—infected tomato plants, the rate of flow may be reduced to 1% of that of healthy plants (Dimond, 1970). It should be noted that resistance to flow in stems of dicotyledons contributes only a very small part of total resistance to flow above ground level. In *Fusarium*-infected tomato plants prior to the onset of visible wilt, stem resistance increased about 500-fold but this only doubled the total above-ground resistance because the major sources of resistance, petrioles and leaves, remained unchanged (Duniway, 1973). Stem resistance remained high after leaves wilted, but more important, petiole resistance approached infinity.

Two kinds of factors are thought to contribute to increased resistance to flow in vascular wilt diseases (Dimond, 1970). First, the pathogen itself, or together with tyloses and vascular gels, may cause partial or complete occlusion of large vessels in the stem. However, since stem resistance is low, these factors alone will rarely reduce flow enough to cause wilting. Additional factors are high molecular-weight polysaccharides produced by the pathogen or cleaved from vessel walls by hydrolytic enzymes. These clog the ultrafilters of pit membranes, restricting lateral transport, and plug the small vessels in petioles and leaves. Although this general concept has much support, Duniway (1973) believes that, except for *Fusarium* wilt of tomato and possibly oak wilt, resistances to water flow which have been reported in plants with vascular wilt diseases are not large enough to account for wilting.

The hypothesis that pathogen-produced metabolites termed wilt toxins play a major role in vascular wilt diseases has been supported chiefly by Gäumann (1958) and his associates. These workers have isolated and characterized a number of toxic compounds capable of disrupting leaf cell permeability from cultures of Fusaria and other wilt pathogens (Kern, 1972). Some of these, lycomarasmin for example, have never been detected in diseased plants whereas others, notably fusaric acid and marticin, have been isolated from infected plants in quantities sufficient to account for wilting and other disease symptoms. Opponents of the toxin hypothesis point out that individually fusaric acid and other wilt toxins do not accurately reproduce the visible symptoms and physiological changes observed in infected plants. This, of course, does not rule out the possibility that several toxins acting sequentially or in concert may be involved.

Until more critical tests are devised the mechanism responsible for symptoms of vascular wilts will remain in doubt. The two hypothetical mechanisms which have been proposed are not mutually exclusive. It is quite possible that wilting results from the combined effects of vascular occlusion and wilt toxins. It is also possible that in some cases vascular occlusion plays the major role whereas in others, toxins are mainly responsible.

3. Stomatal Dysfunction

Changes in transpiration, often followed by rapid dehydration, are common features of many diseases caused by pathogens which do not normally colonize vascular elements. A number of toxic metabolites produced by these pathogens cause rapid, irreversible wilting when they are introduced into the transpiration stream. Results of short-term tests with victorin, the pathotoxin produced by *Helminthosporium victoriae*, and with *H. maydis* T-toxin are shown in Fig. 16a. Each toxin caused a rapid decrease in transpiration which dropped to less than 10% of control rates within one hour. The slightly faster response to T-toxin probably can be attributed to the relatively high concentration used. Examination of silicone rubber impressions taken from victorin-treated leaves at the end of the one-hour test showed that all stomates were tightly closed. Epidermal strips from susceptible maize leaves were floated on solutions of T-toxin and stained with cobaltinitrite to localize potassium. At concentrations of 1% and higher, T-toxin blocked the migration of potassium from subsidiary to guard cells that normally occurs in light and is thought to function in stomatal opening (Arntzen et al., 1973).

In long-term experiments with low concentrations of victorin the initial drop in transpiration is followed after five hours by a rapid increase to a rate slightly above that of untreated controls in light (Fig. 16b). Except for a more precipitous initial drop, the same pattern of changes occurs when victorin-treated plants are kept in the dark. Flaccidity can be detected by the sense of touch soon after transpiration begins to rise (*ca* 6 hrs after treatment) and leaves are visibly wilted when transpiration of victorin-treated leaves rises to that of controls. The wilted leaves continue to transpire for at least 10 hrs at rates slightly above controls (Fig. 16b).

Changes in stomates which accompany victorin-induced changes in transpiration are shown in Fig. 17. Three hours after victorin treatment, when transpiration is at a minimum, the stomates are tightly closed but otherwise normal in appearance with starch grains in guard cells but not in other cells of the leaves (Fig. 17b). Ten hours after treatment stomates are abnormally wide open with guard cells depleted of starch. At this time, large quantities of starch are present in chloroplasts of mesophyll cells which normally do not contain starch (Fig. 17c). Direct examination of intact, toxin-wilted leaves showed that 30–40% of the stomates had grossly swollen guard cells and apertures 2–4 times the width of those of control leaves in light (Fig. 17d).

The effects of fusicoccin on stomatal behavior estimated from resistance to transpiration measured by a diffusion porometer are shown in Fig. 16b. After an

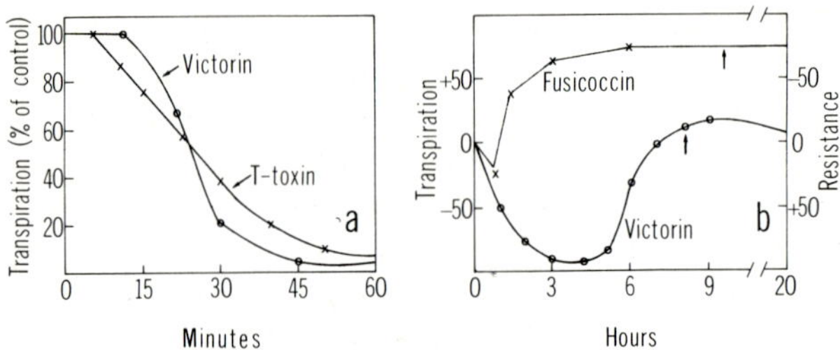

Fig. 16 (a) Short-term effects of toxins produced by *Helminthosporium victoriae* (victorin) and by *H. maydis* (T-toxin) on transpiration of susceptible oat and maize leaves. (b) Patterns of long-term changes in transpiration in oat leaves pretreated with victorin and in epidermal resistance to transpiration in dogwood *(Cornus florida)* cuttings treated with fusicoccin expressed as percentages of controls. Arrows indicate the time at which visible wilting occurred. Data for T-toxin from Arntzen et al. (1973), for fusicoccin from Turner and Graniti (1969) and for victorin from unpublished results from the author's laboratory

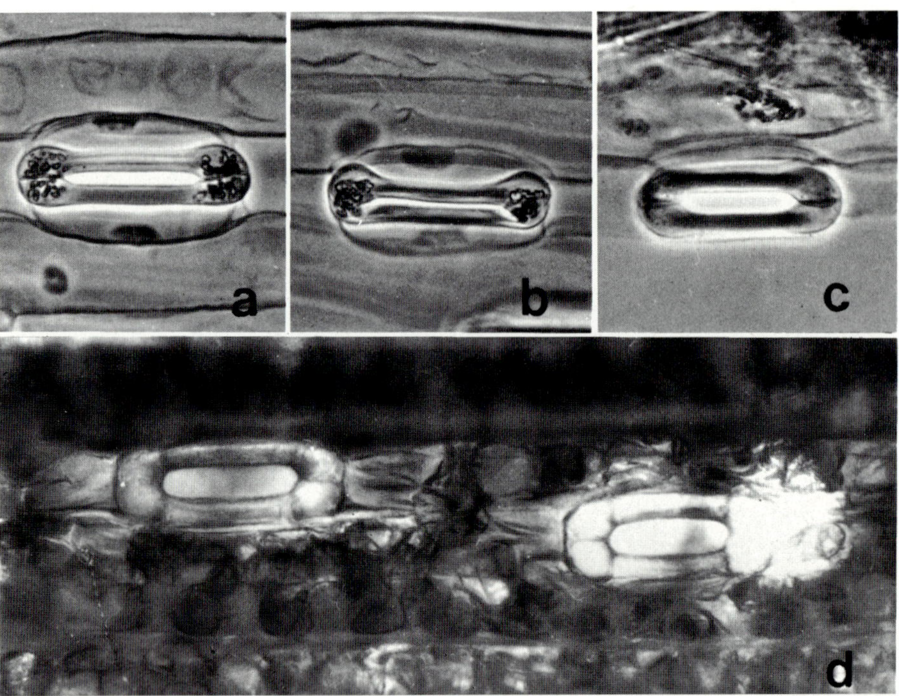

Fig. 17a—d. Effects of victorin, a product of *Helminthosporium victoriae*, on stomates and starch of susceptible oat leaves. (a)–(c) Views of epidermal strips stained for starch with iodine-potassium iodide. (a) Open stomate on an untreated leaf with starch present only in guard cells. (b) Tightly closed stomate three hours after victorin treatment. (c) Abnormally wide-open stomate ten hours after treatment. Starch has disappeared from guard cells but is present in chloroplasts of underlying mesophyll cells. (d) View of intact leaf showing stomates abnormally wide open with swollen guard cells. Photographs by the author

initial increase in resistance which indicated partial transient closure of stomates, resistance fell sharply and remained well below control values even after fusicoccin-treated plants had wilted. Measurements taken from silicone-rubber replicas showed that stomatal widths increased with the drop in resistance. Fusicoccin also causes stomates to open in the dark, and is effective on a wide variety of plant species. Regardless of light conditions, fusicoccin has been reported to stimulate uptake of potassium by guard cells (Turner, 1972) and this result is consistent with evidence that potassium ion accumulation plays a role in stomatal opening.

The mechanisms responsible for stomatal behavior in victorin-treated oat leaves have not been investigated. Since victorin causes large losses of potassium from leaf tissues (Fig. 14b), it is possible that the initial closure is the result of release of potassium from guard cells. The subsequent reopening may be caused by loss of turgor in subsidiary and other cells which surround stomates or by conversion of starch to osmotically active materials in guard cells. The disappearance of starch from guard cells (Fig. 17c) and their highly turgid appearance (Fig. 17d) indicate that the latter mechanism plays a role in keeping stomates open after leaves have wilted. The build-up of starch in mesophyll cells at the same time that starch disappears from guard cells presents a paradox. However, starch often accumulates around lesions produced by leaf pathogens so this effect provides an additional example of similar changes induced by victorin and a variety of other pathogenic agents.

The pattern of transpirational changes induced by victorin (Fig. 16b) is remarkably similar to that observed in rust-infected bean leaves (Durbin, 1967). The secondary increase in transpiration in rusted beans was attributed to an increase in cuticular transpiration caused by ruptures produced by sporulation of the pathogen. Since the increase in transpiration occurred 1 or 2 days before sporulation, it would be worthwhile to determine if stomatal reopening occurs with the rise in transpiration.

III. Effects on Translocation and Mobilization

Abnormal patterns of translocation of organic and inorganic materials are commonly found in plants infected with viruses or biotrophic parasites. Phloem necrosis or gummosis resulting from viral infections obviously could interfere with long-distance transport of photosynthetic products from leaves to tubers or other storage organs. Carbohydrates accumulate in leaves of potatoes infected with leaf-roll virus and there is a corresponding reduction in carbohydrates in the tubers. Phloem necrosis caused by this virus may play a role in these abnormal patterns of carbohydrate accumulation, but starch accumulates in the older leaves before phloem necrosis can be detected. Furthermore, even after phloem necrosis develops in the stems, carbohydrates continue to be rapidly translocated from young leaves. It appears from these results that the leaf-roll virus immobilizes carbohydrates in individual mesophyll cells (Matthews, 1970).

Changes in starch in mesophyll cells surrounding lesions caused by a variety of pathogens have been reported. Mesophyll cells of cereal grasses are normally devoid of starch, and infection, especially by rusts and powdery mildews, results in accumulations of starch similar to those obtained by treating oat leaves with victorin (Fig. 17c). In plants in which starch is present in healthy mesophyll cells, a more complex series of changes has been reported. The starch content of bean *(Phaseolus vulgaris)* leaves declined within 8 hrs after inoculation with rust spores. Three days later the starch content had greatly increased, but seven days after inoculation much of the accumulated starch had disappeared (Mirocha and Zaki, 1966). A similar disappearance of accumulated starch has been reported in late stages of disease development in barley leaves infected with powdery mildew (Bushnell, 1967).

Experiments with radioisotopes indicate that a number of inorganic and organic materials other than carbohydrates accumulate around lesions on diseased plants (Shaw and Samborski, 1956). In addition, there is some evidence of directed, long distance transport of mobile materials from uninfected to infected portions of a plant (Pozsár and Király, 1966). This evidence is not entirely convincing, since most experiments have been of long duration with highly mobile materials such as radiophosphorus. Under such conditions, the observed accumulations may result from increased transport into the infected area, decreased export from the area, or both. Results with ^{45}Ca, which is not readily translocated, indicate that accumulation of this element results form decreased movement away from the infection site rather than an increased movement toward it (Durbin, 1967).

Areas around infection sites, which at least roughly correspond to those in which starch and other materials accumulate, often remain green after other portions of the leaves become yellow. The chlorophyll in these areas, termed green islands, is resistant to removal with organic solvents such as boiling methanol. Green areas which at least superficially resemble green islands can be induced by crude extracts of rust or mildew spores, by various cytokinins, and by siderochromes (natural polyhydroxamate iron-transport compounds). Of the latter, ferrichrome was the most effective; it consistently produced green islands on mature, detached bean leaves at concentrations of 10^{-5} M (Atkin and Neilands, 1972). However, neither cytokinins nor siderochromes produce green islands when applied to intact plants. This indicates that other factors must be involved in green-island formation induced by plant pathogens.

C. Pathological Alterations in Metabolism

I. Changes in Respiration

Changes, usually increases, in respiration are characteristic of most diseased plant tissues. Early investigations with powdery mildews and rusts indicated an initial rise in respiratory activity at or just prior to the development of visible symptoms. Respiration continued to rise with disease development and usually reached a maximum 2 to 4 times higher than that of uninfected tissues at the onset of sporulation. Later the respiratory rate declined. Experiments in which the pathogen was at least partially removed by brushing the leaf surface or stripping off the epidermis indicated that the major portion of the increased respiration was mediated by the plant. This conclusion was strengthened when respiratory increases, similar to those observed with mildewed and rusted tissues, were observed in pathogen-free areas adjacent to lesions produced by nonbiotrophs, in plant tissues infected with certain viruses, and in those treated with pathotoxins.

Some of the hypotheses which have been advanced to account for pathological increases in respiration are discussed in the following sections.

1. The Uncoupling Hypothesis

The hypothesis that pathological increases in respiration result from an uncoupling of phosphorylation from electron transport was advanced by Allen (1953) on the basis of results obtained by Sempio (1950). Sempio's data indicated that ratios of anaerobic to aerobic CO_2 production were markedly lower (ca. 0.3) in diseased than in healthy tissues (ca. 0.5–0.6). This was interpreted by Allen (1953) to represent an abolition of an apparent Pasteur effect in healthy tissues brought about by the uncoupling action of postulated pathogen-produced toxins. Such uncoupling would increase the availability of ADP and thus remove the normal constraint on respiratory activity. It should be noted that ratios of anaerobic to aerobic CO_2 production above 0.33 but below 1.0 are not conclusive evidence of conservation of carbohydrates in the presence of oxygen. Rigorous evidence for the Pasteur effect requires that such conservation be established by appropriate chemical analyses (Beevers, 1961).

Attempts to test the uncoupling hypothesis have yielded mixed results. Rusted wheat leaves failed to respond to the uncoupling agent 2,4-dinitrophenol (DNP) as would be expected if uncoupling had already occurred. These diseased leaves were also insensitive to malonate (a competitive inhibitor of succinic dehydrogenase), but remained as sensitive to fluoride (an inhibitor of enolase in the glycolytic pathway) and to azide (an inhibitor of metal-containing oxidases) as healthy leaves. Homogenates of diseased leaves also showed a marked increase in ascorbic acid oxidation (Farkas and Király, 1955). Remarkably similar results were obtained with susceptible oat leaves treated with the pathotoxin victorin (Krupka, 1959). These data suggested that uncoupling followed by a shift from an iron- to a copper-containing enzyme pathway of oxidation might account for pathological increases in respiration. Also consistent with the uncoupling hypothesis was a report that rusted and DNP-treated wheat leaves contained more inorganic but less organic phosphate than healthy leaves (Pozsár and Király, 1958).

Further work with rusts, victorin-treated oat leaves, and other diseased tissues has cast considerable doubt on the validity of the uncoupling hypothesis. With tissues uncoupled by DNP, the R.Q, or respiratory quotient (ratio of CO_2 evolved to O_2 consumed) rises from a normal value of 1.0 to above 2.0. This is the result of an overstimulation of glycolysis by an excess of ADP to such an extent that oxidation cannot accommodate all the pyruvate produced. The excess pyruvate is shunted to alcohol and CO_2, and the R.Q. increases (Beevers, 1961). No such effect is found with diseased tissues. Instead the R.Q. remains near 1.0 or drifts slightly lower. Uncoupling is not ruled out by these results because the expected rise in R.Q. may be masked by increased dark CO_2 fixation and a shift to oxidation of organic acids.

Results from several laboratories failed to confirm reported increases in inorganic phosphate coupled with decreases in organic phosphate in diseased tissues. Instead, striking increases in organic phosphate were found in infected leaves. Furthermore incorporation of ^{32}P reached a maximum at the peak of the pathological increase in respiration (Heitefuss and Fuchs, 1963). This last finding, in particular, appears incompatible with any extensive uncoupling. Measurements of respiratory control ratios (rate of O_2 uptake without *vs.* with added ADP) and P/O ratios (atoms of P esterified/atoms of O consumed) have also failed to provide direct evidence of uncoupling in diseased tissues. Neither of these ratios changed when victorin was added to well-coupled mitochondria from sensitive oat leaves (Wheeler and Hanchey, 1966). Decreases in these ratios in mitochondrial preparations from leaves pretreated with victorin appeared to be caused by phenolic or other materials released during preparation of the mitochondria.

Increased synthetic activity and, in some cases, increases in growth in diseases caused by biotrophic fungi provide the most convincing evidence against the uncoupling hypothesis. The accumulation of organic and inorganic materials around lesions (see 3. B.III) is energy-dependent since it is blocked by anaerobic conditions, by DNP, and by various metabolic inhibitors. Rust-infected safflower hypocotyls elongate twice as fast as healthy ones and show substantial increases in dry weight (Daly, 1967). This clearly would not be expected if energy generation had been disrupted by uncoupling. On the whole, uncoupling does not appear to be responsible for increases in respiration during early stages of pathogenesis but it may be involved later during tissue degeneration.

2. Involvement of the Pentose Phosphate Pathway

Initial steps in this pathway, which is also known as the direct oxidative pathway or the hexose monophosphate shunt, involve an NADP-dependent oxidation of glucose-6-phosphate followed by decarboxylation at C-1 of the sugar phosphate. These reactions yield ribulose-5-phosphate $+ CO_2 + 2NADPH_2$. Subsequent reactions, which produce 7 and 4 carbon sugar phosphates as intermediates, give rise to phosphorylated hexoses and trioses capable of entering the glycolytic sequence. Alternatively, the hexose product can be converted to glucose-6-phosphate and undergo further oxidation through cyclic operation of the pentose pathway.

In theory, the extent to which the pentose phosphate pathway contributes to the catabolism of glucose can be estimated by short-term measurements of $^{14}CO_2$ evolved from duplicate samples of cells or tissues; one supplied with glucose-6-^{14}C, the other with glucose-1-^{14}C. If catabolism is entirely *via* glycolysis and the TCA cycle, the ratio of $^{14}CO_2$ from glucose-1-^{14}C to that from glucose-6-^{14}C (the C_6/C_1 ratio) should be 1.0. In contrast, if catabolism is entirely *via* the pentose pathway the C_6/C_1 ratio should approach zero in very short-term experiments and then rise with time as the ribulose residues produced by the initial decarboxylation at C-1 are recycled or shunted into the glycolytic sequence. With many fungi, plant pathogenic forms included, initial C_6/C_1 ratios are very low (<0.1) and in some cases rise with time toward unity. In contrast, higher plant tissues usually have ratios in the 0.5–0.7 range and these rise only slightly or not at all with time. Failure to recover equal amounts of CO_2 from C-1 and C-6 even in very long-term experiments invalidates the use of C_6/C_1 ratios for quantitative estimates of the participation of the pentose phosphate pathway in higher plants (Beevers, 1961).

Marked decreases in C_6/C_1 ratios following infection were reported almost simultaneously by Shaw and Samborski (1957) for mildewed and rusted cereal leaves and by Daly and Sayre (1957) for rusted safflower hypocotyls. In all three diseases, ratios dropped from 0.5–0.6 for healthy tissues to 0.2 or lower in advanced stages of disease. In the case of safflower hypocotyls, ratios of anaerobic to aerobic CO_2 production also dropped from 0.62 for healthy tissues (an apparent Pasteur effect) to 0.26 for diseased tissues (a value below the theoretical minimum for the glycolytic-TCA pathway). In addition, diseased safflower tissues were less sensitive to fluoride inhibition than healthy controls. Since the pentose phosphate pathway does not operate anaerobically, CO_2 evolution would be prevented and this would tend to abolish any real or apparent Pasteur effect. This pathway also would not be affected by fluoride which inhibits enolase in the glycolytic sequence. Thus, these results were consistent with a shift from glycolysis and the TCA cycle to the pentose phosphate pathway in diseased plants.

Time-course studies by Shaw and Samborski (1957) indicated that the drop in C_6/C_1 ratios in diseased leaves closely paralleled pathological increases in respiration. In contrast, Daly (1967) and his colleagues reported that respiratory increases occurred at least 24 hrs before C_6/C_1 ratios dropped in rusted safflower, wheat, and bean tissues. Since the decline in C_6/C_1 ratios coincided with the onset of sporulation, Daly (1967) concluded that much of the increased respiratory activity and increased participation of the pentose phosphate pathway was due to metabolic activities of the pathogen. He reported that considerable quantitites of arabitol, a fungal metabolite, could be recovered from diseased leaves whereas only a trace of this compound was detected in healthy ones. Furthermore, the percentage of radioactivity recovered in arabitol from diseased tissues supplied with glucose-6-^{14}C was more than twice that found with glucose-1-^{14}C. In addition to uncertainties about the metabolic contributions of the plant and the pathogen, unexplained fluctuations in respiration and C_6/C_1 ratios and erratic responses to fluoride further complicated interpretation of the role of the pentose phosphate pathway in pathological increases in respiration in plants infected with biotrophic fungi.

Virus-infected plants, tissues adjacent to lesions induced by fungi but free of the pathogen, and pathotoxin-treated tissues provide systems in which changes in plant metabolism during pathogenesis can be studied without interference by metabolic contributions of the pathogen. Symptoms which ranged from necrotic local lesions to mild systemic effects were obtained by inoculating appropriate bean varieties with alfalfa mosaic and southern bean mosaic virus (Bell, 1964).

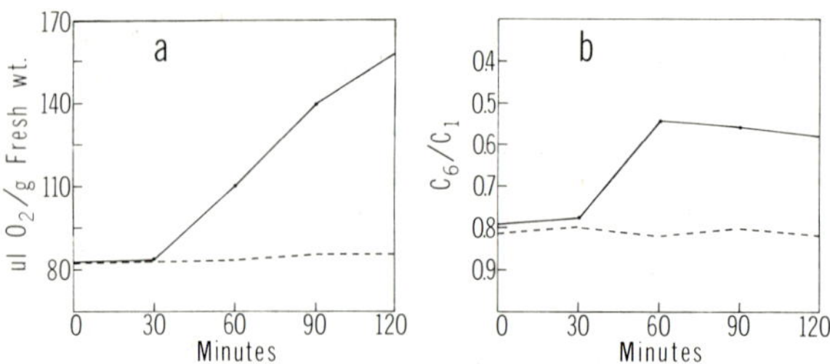

Fig. 18a and b. Time-course of changes in oxygen uptake (a) and in C_6/C_1 ratios (b) in response to victorin applied at time zero. Solid lines, victorin treated; dashed lines, controls. Data from Rawn (1974)

Most of these infections resulted only in small changes in respiratory activity and slight drops in C_6/C_1 ratios. In one local lesion combination, however, respiratory activity increased about 50% and this was accompanied by a drop from 0.5 to 0.3 in the C_6/C_1 ratio. In another study, drops in C_6/C_1 ratios accompanied increases in respiration in tobacco leaves infected with several strains of potato virus X (Dwurazna and Weintraub, 1969). These workers concluded that increased activity of the pentose phosphate pathway was largely responsible for increased respiration. Increased respiratory activity and decreased C_6/C_1 ratios were also found in tissues adjacent to lesions produced by *Rhizoctonia solani* on bean hypocotyls (Bateman and Daly, 1967). Although time-course studies were not attempted, the changes observed in tissues adjacent to young lesions were of the same magnitude as those associated with fully mature lesions.

Pathological increases in respiration, similar in magnitude to those found in rusted and mildewed tissues, can be induced in susceptible oat leaves either by infection with *Helminthosporium victoriae* or by treatment with its pathotoxic product victorin. Victorin-treated oat tissues provide a model system in which metabolic changes characteristic of diseased plants can be induced in a few hours rather than over a period of days. This makes possible quantitative studies of dosage-responses and time-course relationships of pathological changes in metabolism in the absence of any living pathogen.

Time-dependent changes in oxygen uptake and in C_6/C_1 ratios induced by victorin are shown in Fig. 18. No change occurred during the first 30 min of exposure to victorin. During the second 30 min, oxygen consumption rose sharply and this was accompanied by a sharp drop in the C_6/C_1 ratio. Oxygen consumption continued to rise during the next hour whereas the C_6/C_1 ratio remained nearly constant. When tissues were pretreated with victorin for 4 hrs to induce maximal respiratory stimulations (2–3 times control rates) C_6/C_1 ratios dropped and remained at a constant low level for at least 6 hrs. This indicated selective retention by victorin-treated tissues of C-6 of the supplied glucose.

In attempts to determine the fate of the missing C-6, tissues used in C_6/C_1 experiments were fractionated by successive extraction in boiling 80% ethanol followed by hot water, 0.05 N HCl, 10% KOH under nitrogen, and finally by concentrated H_2SO_4. The results showed that C_6/C_1 ratios for the various frac-

Table 5. Distribution of radioactivity in fractions of susceptible oat leaves pretreated with victorin (2 units/ml) or deactivated victorin (controls) and then supplied with specifically labeled glucose

Fraction soluble in	Control[a]			Victorin[a]			% Victorin/Control	
	C_1	C_6	C_6/C_1	C_1	C_6	C_6/C_1	C_1	C_6
Ethanol	101.1	111.0	1.11	115.3	100.2	0.87	115	94
Hot H_2O	5.6	5.8	1.04	6.5	6.6	1.02	116	115
Dilute HCl	4.2	3.7	0.88	8.5	6.8	0.80	203	185
KOH	14.6	12.2	0.84	28.2	28.2	1.00	193	233
Conc. H_2SO_4	15.1	17.2	1.14	37.9	43.1	1.14	250	250

[a] Values are means (DPM $\times 10^{-3}$) obtained with 5 replicates. Differences in C_6/C_1 ratios among treatments were not significant and no C_6/C_1 ratio differed significantly from unity. Data from Rawn (1974).

tions fluctuated around unity with no consistent pattern that might provide a clue as to the fate of the C-6 not recovered as CO_2 (Table 5). In some tests, the total radioactivity recovered was higher for toxin-treated than control leaves as in Table 5 but in others more was recovered from controls or recoveries from treated and control tissues were equal. The only significant difference was the recovery of more radioactivity, both C-1 and C-6, in the combined final three fractions from toxin-treated than from control tissues (Table 5). In a few cases, this did not occur with the dilute HCl extract, which should contain pectic materials, but in all cases the KOH and concentrated H_2SO_4 extracts from toxin-treated tissues contained 2 to 3 times more radioactivity than controls. Presumably the KOH fraction contained primarily hemicellulose and lignin and the H_2SO_4 fraction cellulose. These data indicated that incorporation of both C-6 and C-1 of the labeled glucose into cell-wall materials paralleled respiratory activity which increased 2- to 3-fold in victorin-treated tissues.

Since lignin is hydrolyzed, at least partially, to its phenolic components by alkali, the KOH fraction was acidified and extracted with ethyl acetate. This yielded a phenolic-rich fraction with a C_6/C_1 ratio of about 1.35 for both toxin-treated and control tissues. Such asymmetry, an excess of C-6, would be expected since phenolic components of lignin are synthesized in higher plants primarily *via* the shikimic acid pathway (Neish, 1964). The initial reaction in this pathway involves the condensation of phosphoenolpyruvate and erythrose-4-phosphate. The latter, supplied by the pentose phosphate pathway, would be radioactive if derived from glucose-6-^{14}C but not if derived from glucose-1-^{14}C. The radioactivity from both C-6 and C-1 labels recovered in this fraction was 2 to 3 times greater from toxin-treated than from control tissues. Thus, incorporation into phenolic components of cell walls was also geared to the respiratory rate and this provided additional evidence for a toxin-induced increase in the activity of the pentose phosphate pathway.

Treatment with victorin increased anaerobic CO_2 production nearly two-fold. This indicated a stimulation of glycolysis. At the same time the ratio of anaerobic to aerobic CO_2 production dropped from 0.44 to 0.30.

This abolition of an apparent Pasteur effect, (similar to that observed with rusted and mildewed tissues), would be expected if the activity of the pentose pathway were increased. Treatment of oat tissues with the uncoupler, 2,4-DNP, caused respiratory stimulations of about the same magnitude as those induced with victorin. These, however, occurred immediately (within 15 min) and were accompanied by a rise rather than a fall in C_6/C_1 ratios. Similar results with 2,4-DNP have been obtained with other plant tissues (Shaw and Samborski, 1957; Daly, 1967). Opposite effects on C_6/C_1 ratios of 2,4-DNP and toxin-treatment together with incorporation of labeled glucose into cell-wall components at a rate geared to respiration provide good evidence that victorin-induced increases in respiration do not result from uncoupling.

In general, results with victorin-treated tissues confirm and extend those obtained with plants infected with fungal and viral pathogens. They indicate that an increase in both the glycolytic-TCA sequence and in the pentose phosphate pathway contribute to pathological increases in respiration. Energy consumed in increased synthesis of cell-wall components would increase the supply of ADP and thus remove the normal constraint on glycolysis. The decrease in C_6/C_1 ratio, increase in asymmetrically labeled phenolics, and abolition of an apparent Pasteur effect are all consistent with an increase in the pentose phosphate pathway. Furthermore, the sequestration of asymmetrically labeled phenolics into cell-wall lignins would, at least partially, account for selective retention of C-6 of exogenously supplied glucose. Speculations on how victorin may bring about these changes in respiratory metabolism are reserved for a later section.

II. Effects on Carbon Dioxide Fixation

1. Photosynthetic Activity

Chlorosis resulting from loss of chlorophyll is a common symptom in plants infected with foliar pathogens. Early investigations on photosynthetic activity in cereal grasses infected with rusts and powdery mildews have been surveyed by Sempio (1959). In general, the rate of photosynthesis in infected leaves begins to decline at about the same time that respiration starts to rise. As the disease progresses, photosynthesis continues to decline until eventually the quantity of CO_2 given off in respiration by the plant and pathogen exceeds the amount fixed by the plant. In view of such a net deficit it is not surprising that infection often results in a 50% reduction in total dry weight and even greater reductions in grain yields (Sempio, 1959).

Marked inhibition of $^{14}CO_2$ fixation in the light in late stages of disease has been observed in unifoliate bean leaves heavily infected with rust. At the same time, photosynthetic activity in noninfected trifoliate leaves on the same plants increased to levels 1.5 to 2.0 times that of controls (Daly, 1967). During early stages of infection and prior to inhibition of photosynthesis, large increases in sucrose and smaller increases in glucose and fructose were observed in heavily infected leaves. Later as photosynthesis declined, the quantities of these sugars decreased to about one-half those found in healthy plants.

Changes in starch content following infection have been observed in many foliar diseases. The general pattern is an initial decrease followed by a marked increase with heavy accumulations around the margins of lesions (see 3.B.III). Still later, the starch content again declines. Work with wheat leaves infected with *Puccinia striiformis* led to the suggestion that changes in the activity of ADP-glucose pyrophosphorylase during disease development might account for the pattern of changes in starch content (MacDonald and Strobel, 1970). They assayed this enzyme in the presence of activators (sugar phosphates) and inhibitors (inorganic phosphate) at concentrations found in diseased leaves at various times after inoculation. They reported that the pattern of changes in activity of this enzyme was similar to that of changes in starch content. Changes in starch in leaves infected with *P. striiformis* were very similar to those observed in other diseases. In contrast, photosynthesis in infected leaves increased sharply, reached a peak of twice normal as symptoms appeared, and remained above normal for at least six days after sporulation. These results, which differ markedly from those obtained with other diseases, indicate that *P. striiformis*-infected plants may represent a special case.

2. Dark Carbon Dioxide Fixation

Autoradiographs of leaves exposed to $^{14}CO_2$ in the dark provided the initial evidence for increased dark CO_2 fixation in diseased plants. Plants infected with rust or powdery mildew fixed the greatest amount of CO_2 in the dark at the time of sporulation. Moreover, fixation was concentrated in the area of sporulating lesions. This suggested that much of the increase was due to the ability of the fungus to fix CO_2. Bean and cereal rust uredospores possess the malic enzyme which catalyzes the reversible reaction, pyruvate $+CO_2 \leftrightarrow$malate, in both directions. This enzyme may be responsible at least in part for increased dark CO_2 fixation in diseased leaves (Mirocha and Rick, 1967).

Evidence for increased dark CO_2 fixation in diseased plant tissues in the absence of a living pathogen was provided by Luke and Freeman (1965). They measured tricarboxylic cycle acids (TCA) in victorin-treated oat leaves held in the dark in the presence and absence of CO_2. With CO_2 present, malic and citric acids increased to levels 2 to 3 times above those of untreated control leaves. When CO_2 was removed, citric acid increased but malic did not. No effects on aconitic or succinic acids were observed either in the presence or absence of CO_2. Luke and Freeman (1965) suggested that cations released from the vacuole as the result of victorin-induced disruption of permeability (3.B.I.3) might enhance CO_2 fixation *via* phosphoenolpyruvate to oxalacetate with the latter being converted to malic acid. This mechanism is similar to that proposed to account for increased malic acid synthesis which occurs when large amounts of cations are absorbed. In both cases, increased synthesis of malic acid would serve to buffer the excess cations. Since citric acid accumulation was not dependent on CO_2, its synthesis apparently proceeds primarily through the pyruvate-CoA system.

Further evidence of effects of infection or treatment with pathotoxins on dark CO_2 fixation has been reviewed by Mirocha (1972). The HC-toxin produced by

Helminthosporium carbonum applied to maize causes increases in Dark CO_2 fixation similar to those obtained when oats are treated with victorin. Maize is a highly efficient photosynthetic plant in which the C_4-dicarboxylic acid pathway of CO_2 fixation operates, whereas the major photosynthetic pathway in oats is the Calvin cycle. Both of these pathways use enzymes involved in the synthesis of ribulose-5-phosphate and ribulose-1,5-diphosphate. Mirocha (1972) suggests that both toxins may act by increasing the turnover of these sugar phosphates, thereby increasing rates of CO_2 fixation.

III. Nucleic Acids and Proteins

In many diseased tissues, especially in those characterized by overgrowths, galls or other growth abnormalities, increased metabolic activity is accompanied by marked increases in nucleic acid and protein synthesis. Cytochemical evidence for large changes in nuclei and nucleoli and electron microscopic evidence for changes in ribosomes have been discussed in 3.A.I.1–2. Most additional information has come from work with plants infected with biotrophic fungi or viruses.

1. Nucleic Acid Metabolism

Results from a number of laboratories on effects of infection on nucleic acid metabolism have been reviewed by Heitefuss (1966). Infection of susceptible plants with rust or powdery mildew fungi usually results in an increase in RNA which begins with and parallels the rise in respiration. No such increase in RNA occurs in resistant reactions, and little or no change in DNA is observed in either susceptible or resistant tissues. No consistent pattern of changes in the base composition of RNA was found when healthy and infected susceptible tissues were compared. The changes that were found in base ratios may well have resulted from contamination of plant RNA by that of the fungus during the extraction process.

Heitefuss (1966) suggests that the higher amount of RNA in infected plants is due in part to a delay in the normal decline in RNA that occurs with age. In addition, fungal RNA must contribute to some extent to the total recovered from infected leaves. Despite these uncertainties, the possibility that increased synthesis or exchange of mRNA between plant and pathogen may play an important role in pathogenesis has received some attention. This idea is similar to the hypothesis that the transforming agent in the crown gall disease may be a portion of the bacterial genome which becomes incorporated into host nuclei (Chapter 2.II.1). Day (1974) has discussed the possibility of exchanges of nucleic acids between plants and pathogens. He points out that germinating seeds have been reported to take up and integrate bacterial DNA into their own DNA and also that certain bacterial genes have been reported to be transferred to haploid tomato cell lines. Although these results suggest that transfers of nucleic acids between plants and pathogens may be possible, no evidence of such transfers has been forthcoming.

Very large increases in nucleic acid content have been found in gall cells induced by fungi, bacteria, or nematodes. In cabbage clubroot gall cells which contained large plasmodia of the pathogen *Plasmodiophora brassicae*, nucleolar volumes were 30-fold higher and DNA contents 16-fold greater than in noninfected cells. In the same disease, the nucleolus of root hair cells begins to enlarge within 2 hrs after penetration and after 20 hrs its volume has increased 6-fold. The increase in nucleolar volume is accompanied by an increase in nuclear RNA and nonhistone protein, whereas the lysine-rich portion of histone decreases. These changes suggest that normal transcriptional processes of invaded host cells may be altered by the parasite (Williams et al., 1973).

Results of attempts to determine effects of virus infection on host plant nucleic acids are inconsistent and subject to a variety of interpretations (Matthews, 1970). In various virus-host plant combinations, transitory increases, moderate declines, or no significant changes in DNA have been reported. Such variation probably reflects differences among the plants and viruses studied. This explanation does not account for inconsistent data from a single system: tobacco leaves infected with TMV. Here one investigator reported that transfer RNA increased fourfold after infection, but another, using similar methods, found no such increase. Similarly, a reported rapid breakdown in ribosomes following TMV infection was not confirmed. Marked increases in ribosomes have been observed around necrotic areas on local lesion host inoculated with TMV but no such change occurred in systemic infections. Clearly no generalizations can be drawn from the data now available.

2. Changes in Proteins

The total protein content, when both that of the plant and the pathogen are included, usually increases in early stages of infection. When the contribution of the pathogen is taken into account, little change or, in some cases, decreases in plant protein are found in most diseased tissues. Some however, crown gall tumors for example, contain much more protein than could be expected from synthetic activities of the pathogen. In late degenerative stages of disease, large decreases in protein, sometimes accompanied by increases in free amino acids, are often found.

Uritani (1971) has surveyed changes in proteins in diseased tissues and compared them to those induced by mechanical injury. Most of this work has been done with slices of white or sweet potatoes serving as cut-injured tissue. In slices incubated for 24 hrs, protein increased 10 to 30%, RNA increased 50% and respiratory activity was twice that of freshly cut controls. These changes were accompanied by marked increases in the activity of phenylalanine ammonia lyase, polyphenol oxidase, peroxidase, and a number of enzymes involved in glycolysis and the pentose phosphate pathway. During the initial 24 hrs, change changes in slices inoculated with various fungal pathogens were very similar to those in noninoculated slices. Later, however, abnormal metabolites which were absent or present only in trace quantities in noninoculated slices accumulated in those that were infected. The possible role of such abnormal metabolites in resistance to disease is discussed in Chapter 4.

The development of polyacrylamide gel electrophoresis provided a simple and rapid method for detecting multiple forms of single enzymes known as isozymes or isoenzymes. Use of this method revealed marked changes in isoenzyme patterns in diseased tissue even in some cases where no change in total activity of a particular enzyme could be detected.

Changes in peroxidase isoenzymes have been followed in plants infected with bacteria, fungi, viruses, and in those treated with pathotoxins. In all cases, isoenzyme patterns in diseased tissues differed markedly from those in healthy controls. Some investigators reported the appearance of new isoenzymes, whereas others found only changes in intensity in isoenzyme bands. When techniques were varied to obtain maximum development of very faint bands, no new isoenzymes were found in several virus infections (Novacky and Hampton, 1968). They found, however, that many bands barely detectable in healthy plants greatly increased in intensity following infection and suggested that failure to detect weak activity in healthy plants probably accounted for many reports of new isoenzymes following infection.

Changes in activity and isoenzymes of peroxidase have received much attention because, in certain diseases, increased peroxidase activity has been associated with resistance. The role of this enzyme in normal growth and development is not known but it is thought to be involved in the biosynthesis of ethylene and in the oxidation of indole acetic acid (IAA) and phenolic compounds. Since ethylene and IAA are recognized plant hormones and the oxidation products of phenols may be potent inhibitors, effects on any or all of these systems could result in profound changes in metabolism.

Evidence in favor of the view that changes in enzymes, particularly peroxidase, play a key role in disease reactions has been summarized by Stahmann and Demorest (1973). They point out that increased peroxidase activity has been correlated with resistance to a number of fungal, bacterial, and viral pathogens. Also that resistance to the wild-fire disease, induced by injection of heat-killed cells of *Pseudomonas tabaci* into tobacco leaves, has been reported to be accompanied by marked increases in peroxidase activity. Injection of commercial preparations of peroxidase also gave protection against wild-fire. In addition, they cite reports that sweet-potato slices exposed for 2 days to ethylene became resistant to the black-rot pathogen, *Ceratocystis fimbriata*, and at the same time, large increases in peroxidase and polyphenol oxidase activity occurred. Finally, Stahmann and Demorest (1973) have attempted to test the hypothesis that peroxidase alters histone in such a way that it becomes a less effective repressor of DNA-dependent RNA synthesis. Very preliminary results with calf-thymus histone treated with peroxidase plus H_2O_2 and catechol suggest that alterations in histone do occur in this system. Much more evidence is needed to provide a critical test of this intriguing hypothesis.

Evidence for an important role for peroxidase in disease resistance is far from conclusive and opinion on how the available data should be interpreted is sharply divided. Failure in an independent investigation to confirm the reported effects of ethylene in the sweet-potato—*Ceratocystis* system has been noted in Chapter 2.B.II.3. In a number of other plant-pathogen combinations, ethylene did not induce resistance or produce changes in peroxidase activity. In one case wheat plants which carried a temperature-sensitive gene for resistance were rendered susceptible to rust by exposure to ethylene (Daly, 1972). In the same system, ethylene-induced increases in peroxidase did not render susceptible plants resistant. Furthermore, no change in peroxidase activity occurred when plants resistant at 20° C were transferred to 26° C at which temperature they were highly susceptible. These, and other results reviewed by Sequeira (1973) and Hislop et al. (1973), indicate that the role of changes in peroxidase and other enzymes following infection remains undefined.

IV. Phenol Metabolism

In their discussion of pathological darkening of plant tissues, Rubin and Artsikhovskaya (1964) stress the diversity of physiological roles played by phenolic compounds which characteristically accumulate in diseased plants. Phenolic compounds function as hydrogen donors or acceptors in oxidation-reduction reactions and they play an essential role in lignification. Some phenolics show strong antiauxin activity and are potent growth inhibitors. Others act synergistically with auxins in growth stimulation. Phenolics may also interfere with growth and other energy-dependent activities by uncoupling oxidative phosphorylation. Oxidation of phenols yields highly reactive quinones which inhibit enzymes by complexing with metal ions, reacting with sulfhydryl groups, or by binding nonspecifically to proteins. It is obvious that even small changes in phenol metabolism may severely disrupt many processes which are essential for normal plant growth and development.

Evidence from ^{14}C-tracer experiments indicates that biosynthesis of phenolic compounds in plants may proceed *via* one or more of three different pathways. The most important of these appears to be the shikimic acid pathway in which the initial reaction involves the condensation of phosphoenol pyruvate and erythrose-4-phosphate to yield a seven-carbon intermediate which is converted by a series of reactions into shikimic acid. The patterns obtained with specifically labeled precursors indicate that phosphoenol pyruvate is supplied by glycolysis whereas erythrose-4-phosphate is derived from the pentose phosphate pathway. One reaction, the reduction of 5-dehydroshikimic to shikimic, requires NADPH which is also a product of the pentose phosphate pathway. Transaminations convert shikimic acid to phenylalamine or tyrosine. In dicotyledons, phenylalanine ammonia-lyase (PAL) converts phenylalanine to *trans*-cinnamic acid and thus provides phenylpropane skeletons which can serve as building blocks for lignin, or be utilized in the synthesis of flavanoids and various other phenolic derivatives. In cereal grasses, tyrosine ammonia-lyase converts tyrosine to *p*-coumaric acid (4-hydroxy-cinnamic) and this, in addition to the PAL system, provides phenylpropane skeletons which serve the same functions as those produced in dicotyledons. In several plant tissues, ammonia lyase activity has been found to be closely related to the biosynthesis of phenolic compounds (Kosuge, 1969).

Some phenolic compounds are synthesized by head-to-tail condensation of acetate units. This pathway, known as the acetate-malonate pathway, is thought to involve acetyl Co-A and malonyl Co-A as intermediates and to resemble that of fatty acid synthesis. Isocoumarin (3-methyl-6-methoxy-8-hydroxy-3,4-dihydroisocoumarin) which accumulates in carrot slices is apparently synthesized *via* the acetate-malonate system.

In a third pathway, the acetate-mevalonate pathway, acetyl Co-A units are condensed to form mevalonic acid which is then converted into cyclic terpenes or steroids. The synthesis of compounds with two or more aromatic rings often involves interactions among different pathways. For example, both the acetate-malonate and the shikimic acid pathways contribute to the synthesis of the isoflavone, pisatin.

Major pathways for the biosynthesis of phenolic compounds in plants have been reviewed in detail by Neish (1964) and more briefly by Kosuge (1969). The latter has also discussed various regulatory mechanisms which may modulate different metabolic pathways in diseased plants. Increases in the pentose phosphate pathway observed in infected tissues (3.C.I.2) would yield erythrose phosphate and NADPH and be expected to have a direct influence on the shikimic acid pathway which requires both of these compounds. Reciprocally, NADP produced by both the shikimic and the acetate pathways would serve to maintain high activity of the pentose phosphate pathway. Increased glycolytic activity, which occurs in some diseased tissues, would increase the availability of acetate utilized in acetate pathways of phenol synthesis.

Evidence that two other mechanisms (a) feedback inhibition or activation of enzymes by intermediates or end-products and (b) repression or induction of enzyme synthesis, function in the response of plants to infection is also discussed by Kosuge (1969). A key enzyme in the shikimate pathway, phenylalanine ammonia-lyase (PAL), is strongly inhibited by *trans*-cinnamate and *p*-coumarate. Dehydroshikimate reductase and the coupled oxidation of NADPH is inhibited by cinnamate, catechol, ferulic acid, and other phenolic compounds. These are only two of many enzyme-catalyzed reactions which may control the rate of biosynthesis of phenolic compounds.

In a number of plant tissues, PAL activity is greatly increased by injury or infection. In some cases, these increases in lyase activity can be largely or completely suppressed by low concentrations of puromycin, cycloheximide, or other inhibitors of protein synthesis. These results suggest that synthesis of PAL in uninjured tissues is strongly repressed and that increased synthesis following injury or infection involves derepression. Increased PAL activity induced by injury or infection usually reaches a peak after 24 to 48 hrs and then declines sharply. In bean pods and potato slices, the decline can be prevented if inhibitors of protein synthesis are applied just before the decline would otherwise occur. This suggests derepression of another enzyme system, possibly a protease, capable of destroying lyase and provides another mechanism for regulation of phenolic metabolism (Kosuge, 1969). The hypothesis that derepression plays a key role in disease resistance is discussed in Chapter 5.

Changes in the activity of phenol-oxidizing enzymes, peroxidase and phenolase, may also play a role in the regulation of metabolic pathways in diseased or injured tissues. The controversial nature of evidence that peroxidase plays a key role in resistance was discussed in the preceding section. Phenolase is a copper-containing enzyme which catalyzes the oxidation of mono- and ortho-dihydroxyphenols to quinones. Many reports of increased phenolase activity in extracts of diseased or injured tissues have appeared, but evidence for increased activity of this enzyme *in vivo* has been obtained only with virus-infected tobacco (Hampton, 1970). In many plants, phenolase exists in latent or bound forms, hence activation or solubilization rather than new synthesis may account for increased activity in diseased tissues (Kosuge, 1969). During reactions with ortho-diphenols, phenolase is rapidly inactivated and this makes accurate measurements of its activity difficult. In general, phenolase activity has not been found in extracts of cereal grasses. Instead, large increases in ascorbate oxidation are often observed in diseased cereal tissues but the enzyme involved has not been identified. In addition to ascorbic acid oxidase, both phenolase and cytochrome oxidase are capable of ascorbate oxidation (Beevers, 1961).

Current interest in phenol metabolism stems from the identification of a large number of compounds synthesized *via* the shikimic acid or acetate pathways which accumulate following infection or injury. Many of these compounds, termed phytoalexins, have antibiotic properties and are thought to function in disease resistance. These are discussed in the following chapter.

Chapter 4 Disease-Resistance Mechanisms

In general, disease resistance in plants depends on the ability to tolerate a pathogenic agent or to restrict its development. Tolerance as one type of disease resistance was discussed in Chapter 1.C, and as a factor in insensitivity to selective pathotoxins in Chapter 2.B.III.3. In some cases, preformed, physical or chemical barriers prevent potential pathogens from becoming established (Chapter 2.A.II.3). It is also possible that papillae and cell-wall modifications formed in response to infection retard or prevent some pathogens from penetrating or becoming established within the plant (Chapter 3.A.II.1). In a few cases, preformed toxic materials or deficiency of a nutrient required by the pathogen appear to confer resistance. Very often, however, penetration and initial development of the pathogen take place as readily in resistant tissues as in those that are susceptible. In such cases, disease reactions apparently depend on the type of response induced by the pathogen.

A. Induced Changes in Disease Reactions

Changes in disease reaction from susceptible to resistant or *vice-versa* can be induced by pretreating plants with a variety of physical, chemical, or biological agents. The direction of the change induced by physical and chemical agents often varies with the plant-pathogen combination involved. The effects of ethylene, discussed in Chapter 3.C.III.2 and in more detail by Abeles (1973), provide one example of an agent that has been reported to induce resistance to some pathogens and susceptibility to others. More consistent results have been obtained with plants inoculated first with a potential pathogen to which they are resistant and later with one to which they are normally susceptible. Experiments with a number of plant-pathogen combinations indicate that the initial inoculation renders the plant resistant to subsequent challenge by a normally virulent form. Such induced resistance has attracted much attention because it resembles, at least superficially, immunization in animals. A few examples of induced resistance to viruses, bacteria, and fungi are discussed in the following sections.

I. Induced Resistance to Viruses

Plants infected with certain viruses sometimes apparently recover from initial acute symptoms and become resistant to reinoculation. Tissues surrounding necrotic local lesions, produced as a hypersensitive response to virus infection, also often become resistant to reinfection. These phenomena, which have been known for many years, have been reviewed and discussed by Loebenstein (1972). He points out that although induced resistance in virus-infected hypersensitive tissues is most pronounced in areas around lesions it is also expressed in more distant tissues. Inoculation of half-leaves induces resistance in opposite halves and inoculation of lower leaves on a plant renders upper noninoculated leaves resistant to subsequent challenge inoculation. Induced resistance is not specific for the virus used as the inducing agent. Plants inoculated with one virus have been found to be resistant to several other viruses.

Resistance to viruses has also been induced by fungal infections, by injection of heat-killed bacteria or extracts of bacteria, by infiltration of foreign nucleic acids, and by pretreatment with the pathotoxin victorin. Resistance induced by viral and fungal infections has been associated with necrotic reactions caused by the inducing agent. With the other inducing agents which do not cause visible necrosis, resistance is localized in those tissues directly exposed to the inducer.

In some respects, induced resistance to viruses resembles the interferon system in animals. Interferon is a protein, the synthesis of which can be induced in animal tissues by viable or inactivated viruses and by a number of nonviral agents. Once induced, interferon inhibits the synthesis of new viral nucleic acids. It is possible that induced resistance to plant viruses involves activation of the synthesis of interferon-like substances, perhaps through derepression. To date, however, direct evidence for interferon-like substances in plants is lacking.

Induced resistance has been observed only in plant-virus combinations that result in necrotic local lesions. Such hypersensitive reactions themselves are considered highly efficient resistance mechanisms. The effect of the inducer is to reduce both the size and number of necrotic lesions. Thus, so-called induced resistance to viruses can also be viewed as a suppression of the hypersensitive response.

II. Induced Resistance to Bacteria

Inoculation of plants with avirulent strains of bacterial pathogens or with nonpathogenic bacteria has been reported to protect plants against bacterial diseases. Similar protection has also been obtained with heat-killed bacteria and with cell-free extracts of pathogens or nonpathogens. The fire blight pathogen, *Erwinia amylovora*, which attacks apples, pears, and other members of the Rosaceae, has been studied by several investigators. Goodman (1967) reported that avirulent strains of *E. amylovora* and three other bacterial species protected apple stems from fire blight. Very similar results were obtained with etiolated pear shoots by McIntyre (1974) who also found that cell-free preparations obtained by sonic treatment of virulent or avirulent *E. amylovora* cells gave protection. When DNA was extracted, these cell-free preparations lost their potency and all the activity shown prior to extraction could be accounted for in a purified bacterial DNA fraction.

Tobacco leaves, because of the ease with which suspensions of bacteria or other materials can be introduced by injection or infiltration, have been used extensively to investigate induced changes in resistance. The protective effect of heat-killed bacteria against subsequent challenge with the wild-fire pathogen, *Pseudomonas tabaci* was discussed in Chapter 3.C.III. Heat-killed bacteria not only protect against bacteria pathogenic to tobacco, they also suppress the hypersensitive response (HR) which is induced by bacteria which do not attack tobacco. This response, which can be induced with *P. pisi*, *P. syringae*, or a variety of other phytopathogenic bacteria, is highly reproducible and very striking. Cellular collapse occurs within 9 hrs, and after 18 hrs leaves, infiltrated with bacteria non-pathogenic to tobacco, are bleached and desiccated. In contrast, leaves infiltrated with a tobacco pathogen, such as *P. tabaci*, show symptoms only after 3 to 5 days. Tobacco leaves which exhibit HR evolve large quantities of ammonia and this led to the conclusion that ammonia might be causally related to HR. This conclusion was withdrawn in the face of evidence that large losses of electrolytes and extensive membrane damage occurred long before ammonia evolution could be detected (Goodman, 1972). Fractionation of heat-killed bacteria indicated that proteinaceous substances, probably glycoproteins associated with cell walls, also inhibited hypersensitive responses (Wacek and Sequeira, 1973). Many other treatments—incubation in the dark or at high temperature, infiltration of various chemicals, and even repeated injections of distilled water (Hanchey, et al., 1974)—have been reported to inhibit HR in tobacco. This suggests that trauma or stress, similar to that postulated to inhibit HR in tissues injured by cutting (Chapter 3.A.I), may be involved in suppression of HR in tobacco leaves infiltrated with incompatible strains of bacteria.

III. Induced Resistance to Fungi

In general, agents which induce resistance to fungal pathogens are similar in nature to those discussed in preceding sections on induced resistance to viruses and bacteria. In many cases, induced resistance is associated with HR or necrotic responses and these are accompanied by the accumulation of abnormal metabolites synthesized *via* the shikimic acid or acetate pathways of phenol metabolism. Since these will be discussed in the following section, only a few of many reported cases of induced resistance to fungal pathogens will be considered here.

Potato cultivars carrying R-genes for resistance to various races of the late-blight pathogen *Phytophora infestans* have been used extensively as a model system for studies of induced resistance. The sequence of changes which occurs in potato tissues inoculated with incompatible and compatible races of the pathogen was discussed in Chapter 3.A.I. Potato slices and tissues of other plants inoculated with an incompatible race respond hypersensitively and become resistant to subsequent challenge with a compatible race. Some investigators have reported that when the sequence is reversed, prior inoculation with a compatible race does not prevent induction of the hypersensitive response to an incompatible race (Metlitskii and Ozeretskovskaya, 1968). Such results suggest that compatible (virulent) races do not destroy hypersensitive resistance mechanisms. It should be noted, however, that others have found that the hypersensitive response can be suppressed if potato slices inoculated with a compatible race are incubated for 12 hrs or longer before they are challenged by an incompatible race (Kuć, 1972).

Although slices of potato tubers or sweet-potato roots offer many advantages for studies of the nature of disease resistance, they represent somewhat artificial systems. Furthermore, injury caused by cutting induces both morphological and biochemical changes which may mask or be confused with responses to infection. In a more natural model system, inoculation of uninjured hypocotyls of beans, *Phaseolus vulgaris*, with compatible and incompatible races of *Colletotrichum lindemuthianum* has confirmed results obtained with sliced storage organs. Prior inoculation with fungi nonpathogenic to beans, as well as incompatible races of *C. lindemuthianum* effectively induces resistance to virulent races of the pathogen (Kuć, 1972).

Resistance to fungal pathogens can also be induced by a variety of chemical and physical agents. Effects of temperature changes on the reaction of certain varieties of wheat to stem rust were described in Chapter 2.A. In at least some cases, the effects of temperature cannot be attributed to inhibition or stimulation of growth of the pathogen. Sensitivity to victorin of certain lines of oats, selected from susceptible varieties, is temperature-dependent (Luke and Wallace, 1969), and Japanese pears can be rendered resistant to the pathotoxin produced by *Alternaria kikuchiana* by brief heat treatments (Otani, et al., 1974). Resistance of tomato plants to Fusarium wilt has been associated with increases in phenolic compounds, and pretreatment with low concentrations of catechol induces a high degree of resistance to this disease (Retig and Chet, 1974).

B. Phytoalexins

The term phytoalexin, coined by Müller and Börger (1940), was originally narrowly defined as a chemical produced only when living plant cells, invaded by a parasite, are undergoing necrobiosis with the further requirement that the chemical be responsible for the death of the parasite. Failure over a period of three decades to identify a chemical which met all of these requirements led to redefinitions. Phytoalexins are now generally defined as antibiotics produced in plant-pathogen interactions or as a response to injury or other physiological stimuli (Kuć, 1972). Most chemicals classed as phytoalexins can be detected in small quantities in healthy, nonstimulated plants. Thus, there is no clear distinction between phytoalexins and phytonicides; the latter being defined as antibiotic substances produced by plants as one factor in resistance in interactions of organisms in bionecroses (Metlitskii and Ozeretskovskaya, 1968). Benzoic acid, produced in response to infection of apples by *Nectria galligena* but not detected in healthy or injured apples, may be one example of a chemical that satisfies the original definition of a phytoalexin (Swinburne, 1973).

The intense interest in phytoalexins as possible factors in disease resistance in plants is reflected by numerous reviews. Historical aspects and early investigations have been covered by Cruickshank (1963). The taxonomic distribution and chemistry of characterized phytoalexins have been reviewed by Ingham (1972). The role of phytoalexins in resistance has been evaluated by Metlitskii and Ozeretskovskaya (1968), Stoessl (1970), Matta (1971), and Kuc (1972). Genetic aspects of phytoalexins have been discussed by Day (1974). These sources can be consulted for details of results obtained with various plant-pathogen combinations, some of which are discussed briefly in the following sections.

I. Storage Organ Slices

Slices of potato tubers and of sweet-potato and carrot roots have been used extensively to study the accumulation of compounds with phytoalexin properties following injury or infection. The sequence of events which occurs when potato slices are infected with compatible or incompatible races of *Phytophthora infestans* was described in Chapter 3.A.I, and the changes in phenolic compounds caused by injury (cutting) were discussed in Chapter 3.C.IV. Chlorogenic and caffeic acids, which are normally found in all parts of healthy potato plants, and a large number of abnormal metabolites accumulate in response to infection or injury. Changes in these compounds detected by thin-layer chromatography can be followed as disease progresses (Fig. 19a). Fungitoxic components can be detected by coating the thin layer plates with an agar medium seeded with spores of an appropriate test organism (Fig. 19b).

Total phenols and terpenoids, especially rishitin, accumulate in potato slices inoculated with incompatible races of *P. infestans*. Rishitin, a bicyclic norsesquiterpene alcohol, could not be detected in fresh potato tubers and only traces of this compound were found in aged slices and in those inoculated with compatible races of *P. infestans*. In incompatible reactions, rishitin can be detected at about the same time that growth of intracellular hyphae is inhibited. Fungi not pathogenic to potato and cell-free homogenates of compatible or incompatible races of *P. infestans* induce resistance to virulent races of this pathogen and also the accumulation of rishitin and another terpenoid, phytuberin (Kuć, 1972). Furthermore, suppression of the hypersensitive response in slices inoculated with a compatible race and incubated 12 hrs or longer before challenge with an incompatible one (4.A.III) is accompanied by suppression of necrosis and the accumulation of rishitin and phytuberin.

Large amounts of two other isoprenoids, the steroid-glycoalkaloids, α-solanine and α-chaconine, accumulate in the cut surfaces of potato slices. This accumulation is largely suppressed by inoculation with a compatible race of *P. infestans* and is almost completely suppressed by incompatible races. At the same time, rishitin accumulation is also suppressed in compatible reactions, whereas this compound accumulates to very high quantities in incompatible combinations. This suggests that in incompatible reactions, synthesis is shifted from steroid-glycoalkaloids to rishitin, whereas synthesis of both types of compounds is inhibited in compatible reactions. Such effects might be expected since both rishitin and steroid-glycoalkaloids are synthesized *via* the acetate-mevalonate pathway (Chapter 3.C.IV).

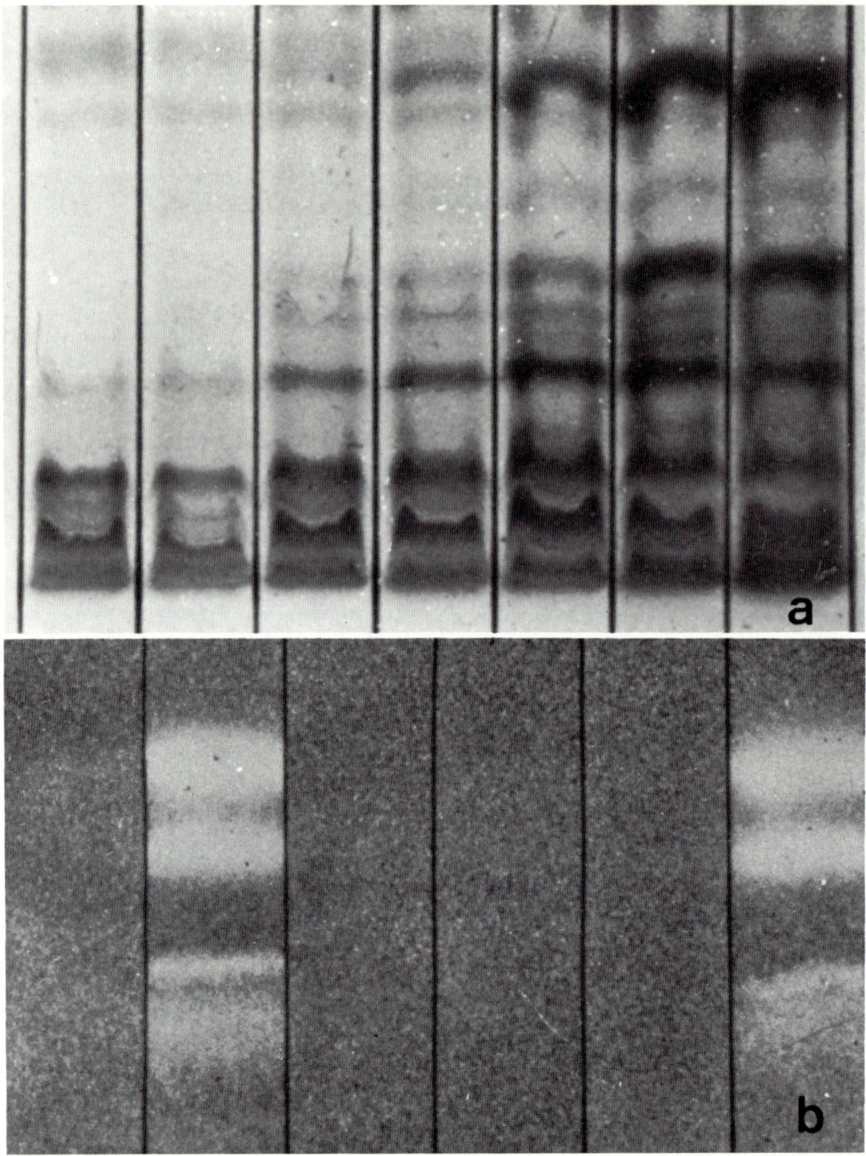

Fig. 19a and b. (a) Thin-layer chromatographs of extracts of potato slices at various times after inoculation with an incompatible race of *Phytophthora infestans* showing a build-up of phenolic and related compounds. From the left, the strips show results at 20, 29, 37, 45, 53, 65, and 94 hrs after inoculation. (b) Zones of inhibition (light areas) obtained when thin layer strips were sprayed with a nutrient solution seeded with spores of *Cladosporium cucumerinum*. The two strips with inhibition zones contain extracts of potatoes infected with in incompatible race of *P. infestans*. The others are various types of controls. Courtesy of J. Kuć

Despite identification of a large number of compounds with phytoalexin properties, the role which these play, individually or in concert, in resistance to *P. infestans* remains unclear. Neither rishitin nor phytuberin could be detected in potato leaves inoculated with virulent or avirulent races of the late blight pathogen. Furthermore, potato leaves of both resistant and susceptible varieties contain fungitoxic steroid-glycoalkaloids in quantities sufficient to inhibit the growth of *P. infestans* completely. Thus, the mere presence of high concentration of phytoalexin-like compounds does not confer resistance. Kuć (1972) has suggested that resistance depends upon some initial event which triggers metabolic alterations including phytoalexin accumulation. Metlitskii and Ozeretskovskaya (1968) suggest that phenols or their oxidation products act synergistically with rishitin and other phytoalexins in resistant reactions. They point out that the fungitoxicity of individual phytoalexins is not very high and that parasites would be expected to adapt rather quickly if only one compound were involved.

Many fungitoxic compounds accumulate in carrot or sweet-potato roots in response to infection or injury. A coumarin derivative 6-methoxymellein (MM), accumulates to concentrations as high as 10^{-2} molal in the peel of intact carrot roots inoculated with *Ceratocystis fimbriata*. Accumulation of MM is thought to be a response to stress rather than a specific response to infection. Accumulation of this compound can be induced by mechanical injury, low temperature, or treatment with ethylene or other chemicals. Kuć (1972) concludes that while MM is not the only determinant of disease resistance, this and other fungitoxic compounds are a part of the defense mechanism in carrot roots. Ipomeamarone and related furano-terpenoids accumulate in *C. fimbriata* infected sweet-potato roots but the role that these play in resistance is not clear. Several investigators have stressed the fact that morphological wound responses, such as periderm formation, complicate interpretation of results obtained with slices of tubers or storage organs.

II. Legume Pods and Hypocotyls

In contrast to the wide variety of fungitoxic compounds associated with disease resistance in roots and tubers, nearly all phytoalexins produced by legumes have in common a pterocarpan structure. Included in this structure are two aromatic rings, one of which is apparently synthesized *via* the shikimate pathway and the other through the acetate-malonate pathway (Kosuge, 1969). (See also Chapter 3.C.IV.) The technique used to isolate pisatin (3-hydroxy-7-methoxy-4′,5′-methylenedioxy-chromanocoumaran) involved inoculation of the endocarp of open pea *(Pisum sativum)* pods with drops of spore suspensions. Materials diffusing into the droplets were extracted and characterized. Similar procedures led to the identification of two closely related compounds; phaseollin produced by beans *(Phaseolus vulgaris)* and 6 α-hydroxyphaseollin produced by soybeans *(Glycine max)*.

Pisatin, phaseollin, and related compounds are weak, broad-spectrum antibiotics produced in response to pathogenic and nonpathogenic fungi and a variety of other stimuli (fungal culture filtrates, ethylene, ultra-violet irradiation, heavy metal ions, and various metabolic inhibitors). Cruickshank (1963) and his colleagues provided the initial evidence that pisatin functions as a phytoalexin. In general, in pods inoculated with nonpathogens of peas, pisatin accumulated to inhibitory levels, whereas in those inoculated with pea pathogens the amounts of pisatin formed were insufficient to inhibit growth of the pathogen. They stressed that neither the amount of pisatin formed nor sensitivity to pisatin alone could account for resistance. Instead, interactions among three factors, the rapidity of the response, the sensitivity of the potential pathogen, and the amount of phytoalexin produced, determine whether a susceptible or resistant reaction occurs.

One possible objection to the use of detached, open legume pods to study phytoalexin production is that they are not the tissues normally attacked and thus represent a somewhat artificial system. In at least one case this objection appears valid. *Phytophthora megasperma* var. *sojae* attacks hypocotyls but not roots or pods of soybean plants. Hypocotyls of resistant plants inoculated with this pathogen produced 10 times more phytoalexin (6 α-hydroxyphaseollin) than those of susceptible plants similarly inoculated. On the other hand, roots, pods, and callus tissues of resistant and susceptible plants produced similar amounts of phytoalexins in response to inoculation (Keen and Horsch, 1972). It is important, therefore, that results with bean pods which led to the identification of phaseollin as a phytoalexin have been confirmed with a more natural system; bean hypocotyls inoculated with various races of *Colletotrichum lindemuthianum* (Kuć, 1972). The bean hypocotyl system has also provided evidence that some factor involved in resistance may be translocated. Bean hypocotyls inoculated with a small drop of a spore suspension of an incompatible race of *C. lindemuthianum*, became resistant to subsequent challenge with a compatible race in an area extending about 1 cm above and below the initial point of inoculation. Neither the fungus nor any accumulation of phytoalexins could be detected in the protected area. This indicated that the translocated factor altered the ability of the protected tissue to respond to the pathogen.

In some cases, virulence may depend on the ability of a pathogen to degrade phytoalexins. Spores or culture filtrates of *Stemphylium botryosum*, a pathogen, or of *Helminthosporium turcicum*, a nonpathogen of alfalfa *(Medicago sativa)* stimulate phytoalexin accumulation in alfalfa leaves. The pathogen, however, degrades the phytoalexin into nonfungitoxic compounds, whereas the nonpathogen does not (Higgins and Millar, 1969). In another interesting system, extracts of broad bean *(Vicia faba)* pollen have been reported to render spores of *Botrytis cinerea* resistant to a fungitoxic compound produced by infected bean tissues, wyerone acid (Mansfield and Deverall, 1971). This may account for increased disease severity observed when pollen is present in the field.

III. Evaluation of the Phytoalexin Theory

Evidence that phytoalexins play an important role in disease resistance can be summarized as follows:

1. In many plants, compounds with phytoalexin properties accumulate as a response to infection or injury.

2. Chemical, physical, and biotic agents which induce resistance to disease also stimulate the synthesis of phytoalexins.
3. In general, phytoalexins accumulate more rapidly and in greater quantities in resistant than in susceptible reactions.
4. In general, nonpathogens are more sensitive to phytoalexins than pathogens.
5. Some exceptions to the general rules in 3 and 4 can be explained by the ability of certain pathogens to degrade phytoalexins to nontoxic products, whereas nonpathogens lack this ability.

Proponents of the phytoalexin theory, currently a vocal majority, find this evidence convincing despite the fact that it is circumstantial rather than direct and conclusive. Some optimistically predict that phytoalexin induction may provide a new method for controlling plant diseases. Skeptics argue that phytoalexins are not produced in amounts sufficient to inhibit pathogens at the time resistance is expressed (Daly, 1972). Proponents reply that even small quantities of phytoalexins may be effective if they are concentrated in a few cells invaded, or adjacent to those invaded, by a pathogen. A more serious problem is posed by results that appear incompatible with the phytoalexin theory. In peas infected with *Aphanomyces euteiches*, lesions expanded rapidly for a period of 5 days even though the quantity of pisatin accumulated after 36 hrs was eight times more than that required to inhibit the growth of the pathogen completely (Pueppke and Van Etten, 1974).

During the past decade, hundreds of papers on phytoalexins have been published. So much investigation indicates that the phytoalexin theory contains at least a spark of truth, but we have as yet only the smoke that must be fanned into flame by compelling new evidence.

C. Lysosomes

Lysosomes are membrane-bound subcellular organelles rich in acid hydrolases. Originally associated with autolysis of animal cells they were termed "suicide bags." Currently they are considered to be important both in normal animal metabolism and in a number of pathological conditions (deDuve, 1969). In plants, acid hydrolases have been found not only in lysosome-like particles in the cytoplasm but also in vacuoles, cell walls, and in nuclei. Primary lysosomes may be synthesized by the Golgi apparatus and later fuse with vacuoles which are thought to function as secondary lysosomes in plant cells (Wilson, 1973).

The fragmentary evidence that lysosomes play a role in disease reactions has been discussed by Pitt and Galpin (1973) and by Wilson (1973). The former cite histochemical evidence that hydrolytic enzymes are released from lysosomes during infection of potato tubers by fungal pathogens. They also report increases in the activity of several lysosomal enzymes as a response to infection. Wilson (1973) suggests that disruption of lysosomal membranes and release of lysosomal enzymes may be responsible for the necrosis which occurs in hypersensitive responses. Such a mechanism would also account for the sudden granulation of protoplasm which is often observed at the time of cell death (Chapter 3.A.I).

Chapter 5 Genetics of Pathogenesis

Events which occur during pathogenesis reflect the combined activities of the genetic systems of the plant and the pathogen. In a discussion of these genetic interactions, Day (1974) points out that our limited knowledge of gene function in pathogenesis requires a speculative approach. This approach, which is taken here, may be justified if it leads to concepts subject to experimental verification.

A. Genetic Dissection

If pathogenesis involves an ordered sequence of events, it should be possible to analyze the process by the use of spontaneous or induced mutants blocked at different steps in the sequence. This technique, aptly termed "genetic dissection" by Levine (1968), has been used with great success in the elucidation of biosynthetic pathways in microorganisms and in analysis of the sexual process in fungi (Wheeler, 1969). Genetic dissection has been used in a preliminary way to identify stages in the infection process at which resistance is expressed in wheat and barley infected with powdery mildew (Ellingboe, 1972). No genes were found which blocked spore germination, appressorial maturation, or the formation of infection pegs. However, in several different incompatible interactions, the percentage of infection pegs that produced rudimentary haustoria was sharply reduced. In some of these interactions, genetic blocks at later stages of the infection process, haustorial maturation and the formation of secondary hyphae, were identified. These results indicate that genetic dissection of pathogenesis offers a feasible approach to the problem of how genes in the plant and pathogen function to determine disease reactions.

B. The Gene-for-Gene Concept

This concept evolved from the combined results of genetic analyses of flax *(Linum usitatissimum)* and the flax rust fungus *(Melampsora lini)*. The concept states that for each gene conditioning avirulence or virulence in the pathogen there is a corresponding gene conditioning resistance or susceptibility in the plant (Flor, 1955). In applying this concept, disease reactions were reduced to only two classes. Reactions rated 0–2 were considered resistant, whereas those rated 3 or 4 were classified susceptible. With this method, extensive data from the *Linum-M.lini* system appeared consistent with the postulated one-to-one relationship between genes in the plant and the pathogen (Flor, 1971).

The importance of a one-to-one relationship, considered by some to be the essence of the gene-for-gene concept, has probably been overemphasized. As Flor (1955) has pointed out, genes in the plant and pathogen are assigned on the basis of a single character; the type of lesion produced by infection. Thus, a one-to-one relationship is virtually assured.

Whether or not a one-to-one relationship holds in all cases is of little importance. What is important is how genes in the plant and pathogen interact to determine disease reactions. Often overlooked is the fact that the gene-for-gene concept assumes that resistance will be expressed in one, and only one interaction; the combination of a resistant plant and an avirulent race of the pathogen. In genetic terminology, if a pair of alleles control resistance in a plant and a corresponding pair govern pathogenicity in a pathogen, a resistant reaction will occur if, and only if, a gene for resistance interacts with a gene for avirulence. This is the rule used by Flor (1955) in his genetic analysis of the *Linum-M.lini* system. In this system, resistance was dominant to susceptibility in the plant and avirulence was dominant to virulence in the pathogen.

Two varieties of flax and two races of the flax rust fungus can be used to illustrate Flor's gene-for-gene system:

Rust race	Genotypes	Flax variety	
		Ottawa 770B L	Bison l
123	A_L	Resistant	Susceptible
108	a_L	Susceptible	Susceptible

Note that with dominance, diploid plants and dicaryotic pathogens can be treated as haploids, i.e., L=LL or Ll in the plant and $A_L = A_L A_L$ or $A_L a_L$ in the pathogen. Also that contrary to accepted definitions, susceptible reactions are assumed to occur in $l + A_L$ (plant susceptible, pathogen avirulent) and in $L + a_L$ (plant resistant, pathogen virulent) interactions as well as in the $l + a_L$ (plant susceptible, pathogen virulent) interaction. Acceptance of these assumptions has led to such self-contradictions as "avirulent pathogenicity," and "virulence" defined as the ability to produce disease on a resistant plant. Such usage has caused much confusion, and, as will be seen later, can and should be avoided.

I. The Quadratic Check

Loegering and Powers (1962) were the first to point out that Flor's system of genetic analysis could be stated in general form as four possible interactions among a pair of alleles governing resistance in a plant and a corresponding pair controlling virulence in the pathogen. The result shown below, called a quadratic check, is that expected on the basis of Flor's gene-for-gene concept.

		Plant genotypes	
		R	r
Pathogen	A	Resistant	Susceptible
Genotypes	a	Susceptible	Susceptible

Note that only a resistant reaction is definitive in that it specifies the genotypes of both plant and pathogen (R + A). By itself, a susceptible reaction eliminates R + A but gives no information as to which of the other three interactions are involved. This follows from Flor's rule, resistance will be expressed if, and only if, a gene for resistance (R) in the plant interacts with a gene for avirulence (A) in the pathogen. In the quadratic check, capital and lower case letters are used merely to differentiate alleles and do not imply dominance or recessiveness.

The revolutionary nature of Flor's gene-for-gene concept becomes apparent when experience in an introductory course in plant pathology is recalled. Very often a somewhat sadistic professor presents his students with a basket of half-rotten fruits and vegetables and tells them to satisfy Koch's postulates. After vain attempts with the saprophytic bacteria and fungi that grow out so luxuriantly from bits of diseased tissue on agar plates, the student learns that to induce disease one must have a susceptible plant and a virulent pathogen. This can be stated as a rule of definition; a susceptible reaction will occur if, and only if, a susceptible plant is infected with a virulent pathogen. Or, in genetic terms, susceptibility will be expressed if, and only if, a plant with a gene for susceptibility (r) interacts with a pathogen with a gene for virulence (a).

Application of the rule of definition can be illustrated by results obtained with the Victoria blight pathogen, *Helminthosporium victoriae*. Subcultures from a single-spore isolate of *H. victoriae* yielded a virulent isolate I and an avirulent isolate II. Mass screening of a susceptible oat variety, Victorgrain 48-93 (Vg 48-93), yielded a *H. victoriae* resistant (HVR) selection presumed to be a mutant since it differed from its putative parent only in disease reaction (Luke et al., 1960). The reactions which occur with this system when the rule of definition is applied are as follows:

		Oat lines	
H. victoria	Genotypes	HVR R	Vg 48–93 r
Isolate I	a	Resistant	Susceptible
Isolate II	A	Resistant	Resistant

It is obvious that results with the Victoria blight system are the opposite of those obtained with flax rust. Susceptibility, which requires a specific interaction of a gene (r) for susceptibility in the plant and one for virulence (a) in the pathogen, is the definitive reaction. Resistance will be expressed in all of the other three possible combinations.

Day (1974) points out that selectively toxic chemicals can be substituted for a pathogen in the quadratic check. For example, in the Victoria blight system, victorin can be used instead of the virulent isolate I which produces it, and deactivated victorin in the place of isolate II, a nonproducer. Day (1974) fails to note, however, that the results are not those predicted by Flor's rule. Instead, they are consistent with the rule of definition.

The question now raised is which, if either, of the two rules is valid. Contrary to statements which have appeared frequently in the literature, this question cannot be settled by the results of quadratic checks or by more extensive genetic analysis such as that carried out with flax rust. Wherever only two classes of interactions are involved, either rule can be applied. Applying Flor's rule to the Victoria blight system we get:

		Oat lines	
H. victoriae	Genotypes	HVR RR_1	Vg 48–93 rR_1
Isolate I	Aa_1	Resistant	Susceptible
Isolate II	AA_1	Resistant	Resistant

All that is needed is the assumption that both plants have an additional gene for resistance and both isolates of the pathogen an additional gene for avirulence. The plant lines and the fungal isolates remain isogenic, thus the presence of additional genes cannot be detected by conventional genetic analysis. Application of Flor's rule to the Victoria blight system brings out a further assumption of the gene-for-gene concept. Resistance will be expressed if any gene for avirulence in the pathogen interacts with a corresponding gene for resistance in the plant no matter how many other interacting pairs that would give susceptible reactions are involved. Thus, when HVR is inoculated with isolate I, the A+R interaction blocks the otherwise susceptible reaction expected from a_1+R_1.

In a similar way, the rule of definition can be applied to results with flax rust.

		Flax variety	
Rust race	Genotypes	Ottawa 770B Rr_1	Bison rr_1
Race 123	aA_1	Resistant	Susceptible
Race 108	aa_1	Susceptible	Susceptible

Again an additional gene pair is required, and in this case, the assumption that any gene pair interacting to condition susceptibility will override the effects of any other interaction.

In applying his rule, Flor (1955) has said only that it was the simplest hypothesis that would account for his results with flax rust. Although this is true for the data from flax rust used here for illustration, it is obviously not true for the Victoria blight data. Furthermore, as more gene pairs are added, neither system has an advantage in simplicity. In addition, rules other than the two which have been discussed can be devised for genetic interactions between plants and pathogens. Favret (1971) has proposed a model in which a pair of alleles in a pathogen interact with a single allele of a multiallelic series in the plant. In this model, genes for virulence induce a susceptible response in the host, whereas genes for avirulence merely fail to activate such a response. Favret (1971) has applied his model to results with powdery mildew of barley and Daly (1972) has discussed susceptibility as an induced response in wheat rust.

At this stage it may appear that genetic analysis of pathogenesis is a game that anyone can play according to his own rules. It should be recalled, however, that Flor's rule evolved from results of a search for resistance to a biotrophic pathogen, whereas the rule of definition emerged as a logical consequence of attempts to induce disease with nonbiotrophs. (Even professors are not sadistic enough to require naive students to satisfy Koch's postulates with a rust fungus). It is therefore possible that two contrasting sets of rules are required; one for biotrophs, which in nature require a living plant to complete their life cycles, and another for nonbiotrophs, which are capable of saprophytic growth and reproduction. In essence, this is the view adopted by many proponents of Flor's gene-for-gene concept. Loegering (1972) has termed the interaction of a plant and a biotrophic pathogen an "aegricorpus;" a unique, two-component structure comparable in many ways to a lichen or a mycorrhiza. He has also proposed that the character of the aegricorpus be described as an infection type (it), either high (Hit) or low (Lit). Plant reactions would be termed low (Lr) or high (Hr) rather than resistant or susceptible, and high pathogenicity (Hp) or low (Lp) would replace virulence and avirulence to describe attributes of the pathogen. He points out that adoption of these conventions would circumvent semantic problems involved in Flor's gene-for-gene concept.

At first glance, the idea that genetic mechanisms which control pathogenesis in diseases caused by biotrophs differ from those in which nonbiotrophs are involved appears attractive. In most diseases caused by biotrophs, low pathogenicity (avirulence) in the pathogen and low reaction (resistance) in the plant are dominant characters. Since dominance is usually associated with active processes (the production of a substance rather than the inability to produce it) these results are those expected from Flor's rule. Also, most mutations observed in biotrophic pathogens have been from avirulence to virulence, and mutations resulting in loss of synthetic ability are more common than those in the opposite direction. It has been suggested that deletions of genes for avirulence may be responsible for mutations in rust fungi since, by Flor's rule, these would produce virulent races.

Very similar arguments can be advanced for application of the rule of definition in diseases caused by nonbiotrophs, especially those in which pathotoxins are involved. In the case of Victoria blight, susceptibility is dominant, and virulence would also be expected to be dominant since it depends, at least in part, on the ability of the pathogen to produce the toxin, victorin. In contrast to results with biotrophs, virulence is commonly lost rather than acquired by mutations in nonbiotrophic pathogens.

Certain facts appear inconsistent with the idea that separate and essentially opposite genetic mechanisms operate in pathogenesis induced by biotrophs and nonbiotrophs. Early, and probably crucial, events in pathogenesis are very similar in plants infected with the two types of pathogens. In susceptible interactions, compatibility is the rule; both biotrophs and nonbiotrophs grow and ramify in susceptible tissues for a considerable time before host responses are evident. In resistant interactions, nonbiotrophs induce the same type of hypersensitive response that occurs with avirulent (incompatible) races of biotrophic pathogens. In later stages of pathogenesis, physiological changes induced by nonbiotrophs are remarkably similar to those elicited by biotrophs. Finally, Flor (1971) has listed several diseases caused by nonbiotrophs among the plant-pathogen systems to which his gene-for-gene concept has been applied.

In summary, results of genetic analysis of a number of plant-pathogen systems are consistent with that part of Flor's concept which states that for each gene conditioning resistance in a plant there is a corresponding gene that conditions pathogenicity in the pathogen. As pointed out earlier, such a relationship is virtually assured by the use of the infection type as a single character for the identification of genes in both organisms. A more fundamental aspect of Flor's gene-for-gene concept holds that a low infection type (resistance) requires the specific interaction of a gene for low pathogenicity (avirulence) in the pathogen and a gene for low reaction (resistance) in the plant. Deletion of either gene would lead to passive susceptibility. The validity of this aspect of Flor's concept has not been established. The available genetic data can be interpreted on the basis of the more conventional assumption that susceptibility is definitive in that it requires a gene for virulence in the pathogen and one for susceptibility in the plant. On this assumption, passive susceptibility is ruled out but deletion of the virulence gene would lead to passive resistance.

Loegering (1972) has expressed the view that Flor's gene-for-gene concept has been of value chiefly as a stimulus for the formulation of hypotheses about the fundamental nature of disease and the mechanism of action of genes which control pathogenesis. A number of such hypotheses and experiments designed to test them are discussed in the following sections.

C. Biochemistry of Pathogen Specificity

The genetic models discussed in the preceding section are based on the assumption that genes in the plant, the pathogen, or both, control the production of substances which determine the nature of disease reactions. Flor's rule predicts that the product of a gene for avirulence in the pathogen is required to elicit a resistant response. In contrast, the rule of definition predicts that a gene for virulence in the pathogen controls the production of a substance required for a susceptible reaction. The rather meager evidence for the existence of gene products of either type is discussed below.

I. The Role of Selective Pathotoxins

The closest approach to the identification of gene products which determine disease reactions and account for pathogen specificity has come from work with selective pathotoxins discussed in Chapter 2.B.III.3. All fungal pathogens which have been shown to produce selective pathotoxins are nonbiotrophs and many belong to a single genus, *Helminthosporium* (Table 3). All selective pathotoxins produced by fungi are low molecular weight substances which at appropriate concentrations cause disease symptoms only on susceptible plants. Loss of toxin-producing ability by the pathogen is accompanied by loss of ability to cause disease. On the surface, at least, these results are those predicted by the rule of definition and quite the opposite of those expected from Flor's concept.

An ingenious hypothesis advanced by Litzenberger (1949) to account for an apparent complete linkage of susceptibility to *Helminthosporium victoriae* and resistance to certain races of crown rust in oats provided the basis for attempts to reconcile results with victorin with Flor's rule. Litzenberger (1949) proposed that a similar toxin was produced by both pathogens. The lethal effects of the toxin on sensitive plants would promote growth of *H. victoriae* and result in susceptibility whereas growth of the biotrophic rust fungus would be inhibited and a resistant reaction of the hypersensitive type would be expected. If reactions to victorin and *H. victoriae* which have been classed susceptible are considered instead to represent hypersensitive (resistant) responses, then the results of the quadratic check with the *H. victoriae* system discussed in the preceding section would be compatible with Flor's rule. In this view, which was adopted by Hadwiger and Schwochau (1969), victorin would be a specific product of a gene for avirulence in the pathogen which activates a massive hypersensitive response in sensitive plants.

Unfortunately, the responses of sensitive plants to victorin or to infection with *H. victoriae* do not appear to be of the hypersensitive type. Sensitive leaves treated with high concentrations of victorin dry out rapidly and appear withered but they do not become necrotic. If low concentrations are used, no visible symptoms are seen for 2 to 3 days after which orange-red streaks appear which are typical of those produced in natural infections. Results of a histological study (Paddock, 1953) indicate that *H. victoriae* grows and ramifies extensively in susceptible tissues before host cells are killed or necrosis occurs. The only typical hypersensitive response to infection with this pathogen occurred in a resistant variety. These results indicate that Flor's gene-for-gene concept cannot account for the response of susceptible tissues to victorin or *H. victoriae*.

It should be noted that the mechanism of gene action postulated by Flor cannot be entirely ruled out in the Victoria blight system. Although victorin appears to be responsible for disease symptoms in susceptible plants, there is no evidence that victorin plays a role in the hypersensitive reactions observed when resistant plants are infected. Instead, resistant tissues appear to be highly insensitive to victorin. It is possible, therefore, that resistance to *H. victoriae*, but not to victorin, depends on a gene product not yet identified which induces a hypersensitive response in resistant plants.

II. The Role of Cell Wall-Degrading Enzymes

Extracellular enzymes, in particular those with pectolytic activity, capable of degrading cell walls are produced by many plant pathogens. In general, such enzymes have not exhibited selective properties. Preparations obtained from pathogens or nonpathogens commonly degrade cell walls of a variety of plants. Certain exceptions to this rule have been reported. With age, bean hypocotyls become resistant to *Rhizoctonia solani* and *Colletotrichum lindemuthianum*, and concomitantly cell walls of these hypocotyls become resistant to degradation by enzymes produced by these pathogens (Albersheim et al., 1969). Initial results with the bean anthracnose system indicated that α-galactosidase was produced in greater quantities when the pathogen was grown on cell walls from susceptible plants than when grown on those from resistant ones. However, further work with this and other systems failed to provide clear evidence for such selectivity (Wood, 1973).

Although a role for cell wall-degrading enzymes in pathogen specificity has not been established, it is clear that they function as mechanisms of attack in many plant diseases (Chapter 2.B.I.2). As gene products associated with susceptibility, they lend support to the rule of definition.

III. Evidence for Specific Inducers of Resistance

Despite extensive searches there is little evidence for materials produced under the control of specific avirulence genes which, on the basis of Flor's concept, should have selective effects on resistant plants. Work in Holland which indicated that filtrates of avirulent cultures of a certain strain of *Venturia inaequalis* were selectively toxic to resistant apple leaves was not confirmed in later investigations (Day, 1974).

In another system, losses of radioactivity from tomato-leaf disks labeled with ^{32}P and then infiltrated with a high molecular weight fraction of a culture filtrate of *Cladosporium fulvum* were followed (Kaars Sijpesteijn and van Dijkman, 1973). In tests involving fractions from several races of the pathogen, plants resistant to a particular race lost more ^{32}P when treated with a fraction from that race than did susceptible plants or untreated controls. These results are of considerable interest since they are clearly in accord with Flor's gene-for-gene concept. They must, however, be regarded as preliminary since, in discussion, one investigator indicated that in some tests results were negative (Kaars Sijpesteijn and van Dijkman, 1973).

IV. The Common Antigen Hypothesis

The idea that plants and pathogens which interact to give susceptible (compatible) reactions have more antigens in common than those that give resistant (incompatible) reactions was explored first with the flax rust system. Tests of globulin antigens from four races of rust and from four lines of flax indicated that susceptibility depended on the presence in the plant of a protein serologically similar to one possessed by the pathogen. These results were questioned on the basis that rust spores might be contaminated by host antigens, but Flor (1971) states that they have been confirmed in tests in which all spores were produced on a single flax variety.

Further attempts to explore the hypothesis that antigens shared by the plant and pathogen may account for specificity have been carried out with several types of pathogens (De Vay et al., 1972). Pathogenic races of *Xanthomonas malvacearum*, which causes angular leaf spot of cotton, had more antigens in common with those of cotton leaves than did nonpathogenic races of bacteria. Similar results were obtained with *Ceratocystis fimbriata* which attacks sweet-potato roots. With both of these systems, positive results were obtained only when fully susceptible interactions were compared with highly resistant ones of the hypersensitive type. In another system, *Fusarium* wilt of cotton, such relationships were not observed. In this case, loss of virulence in the pathogen did not result in a marked change in antigenic relationships between the plant and pathogen. The smut fungus, *Ustilago maydis*, shared an antigen with susceptible maize and showed a close serological relationship to young oat seedlings which were rapidly killed by this fungus. Wood (1973) points out that these results are puzzling since they indicate that common antigens are involved in a compatible relationship on maize and also in a highly incompatible one on oats.

Taken together, serological data suggest that common antigens may play a role in pathogen specificity. They do not, however, provide a basis for support or rejection of any model of gene action in pathogenesis.

V. The Phytoalexin-Induction Hypothesis

Phytoalexins can be viewed on the basis of Flor's gene-for-gene concept as products of the interaction of specific genes for avirulence in pathogens with corresponding genes for resistance in plants. This hypothesis was first advanced in a formal way by Hadwiger and Schwochau (1969) to account for effects of certain microbial metabolites on pisatin synthesis in detached pea pods. They observed that at low concentrations actinomycin D and a variety of other agents stimulated pisatin production whereas high concentrations of the same agents inhibited the production of pisatin. To explain these results, they invoked the Jacob-Monod model of gene action and regulation. In this induction hypothesis, enzymes required for pisatin synthesis are assumed to be coded by a polycistronic gene with an operator site. Normally a regulator gene produces a specific repressor which combines with the operator to keep the operon and pisatin synthesis turned off.

At low concentration, agents such as actinomycin D preferentially complex with the regulator, thus blocking synthesis of the repressor and inducing pisatin synthesis. At high concentrations the same agents inhibit pisatin production by blocking other sites involved in the synthesis of this compound.

Evidence that phenylalanine ammonia lyase (PAL), a key enzyme in the biosynthesis of pisatin, is strongly repressed in healthy plant tissues (Chapter 3.C.IV) provides some support for the phytoalexin-induction hypothesis. Also, all of a large number of different agents (heavy metals, polypeptides, antibiotics, and drugs) which induce pisatin synthesis also cause increases in PAL activity. On the other hand, lack of specificity of the inducing agents and failure to identify any product of a gene for avirulence in a pathogen as an inducing substance leave open the question of the role of inducers of PAL activity in disease reactions (Day, 1974).

Some support for the induction hypothesis was provided by a report that flax rust infection by a virulent race of the pathogen is reduced by prior inoculation with an avirulent race (Littlefield, 1969). Also consistent with this hypothesis is a report of early net increases in protein synthesis in resistant but not in susceptible interactions in the flax rust system (Von Broembsen and Hadwiger, 1972). In contrast to these results is a report of an early loss of polysomes from susceptible but not from resistant barley leaves infected with powdery mildew (Dyer and Scott, 1972).

On the whole, results with phytoalexins discussed in the preceding chapter have not provided clear evidence for any particular model of gene action or regulation in pathogenesis. They have, however, provided evidence that disease reactions can be reversed by prior treatment with pathogens, nonpathogens, subcellular fractions or extracts, and a variety of other agents. Kuć (1972) points out that the potential for resistance appears to exist even in completely susceptible cultivars considered to have no genes for resistance. Since such cultivars can be rendered resistant by a variety of treatments, all that is required is activation for the inherent potential for resistance to be expressed.

VI. Summary

Flor's gene-for-gene concept implies that resistance is an induced response which requires the production of an inducing substance by an active gene for avirulence in the pathogen. If the gene for avirulence is inactive or absent, passive susceptibility will be the result. The alternative concepts discussed in this chapter imply that the induced response is susceptibility which requires an active gene for virulence in the pathogen. Results of genetic analyses of plants and pathogens can be interpreted in terms of either induced resistance or induced susceptibility; thus there is no direct evidence for either mechanism.

Selective pathotoxins as disease-inducing agents produced by virulent pathogens provide indirect evidence that products of genes for virulence are essential for susceptible reactions in certain diseases as predicted by the induced susceptibility concept. Although clear evidence for products of genes for avirulence which induce resistance is lacking, a role for such products in resistance, even in diseases in which selective pathotoxins are involved, cannot be ruled out. It is possible that resistance to pathogens which produce pathotoxins involves two separate mechanisms. First, insensitivity to the toxin which can be considered the product of a gene for virulence, and second, the ability to respond in a hypersensitive fashion to an undetected product of a gene for avirulence.

On the basis of present evidence, Flor's gene-for-gene concept appears ruled out as the sole mechanism responsible for disease reactions where pathotoxins are involved. It is possible that, as originally proposed, Flor's concept applies only to diseases caused by biotrophs. However, the similarities of initial events in pathogenesis induced by biotrophs and nonbiotrophs argue against the involvement of two separate and essentially opposite mechanisms. Until these uncertainties have been resolved, the hypothetical nature of all proposed models of gene action in pathogenesis should be emphasized. Hopefully, those who employ Flor's system will adopt the terminology suggested by Loegering (1972) and thus avoid further semantic confusion.

D. Genetic Vulnerability

Two epidemics, Victoria blight of oats in the late 1940's and southern corn leaf blight in 1970, drew attention to the dangers of genetic uniformity in major crop plants. Each epidemic followed the introduction of a single genetic character into nearly all commercial cultivars of a crop. The result was a population uniformly susceptible to a previously unimportant pathogen. In the case of Victoria blight, the introduced gene, Pc-2, conferred a high degree of resistance to crown rust. In the southern corn leaf blight case, the character was a cytoplasmic factor, Tms, for male sterility. Each of these characters was associated with susceptibility to a species of *Helminthosporium* which produced a selective pathotoxin (see Chapter 2.B.III.3).

The outbreak of southern corn leaf blight led to a study of genetic vulnerability by a committee of the U.S. National Academy of Sciences (1972). In their report, this committee drew two main conclusions: (a) vulnerability stems from genetic uniformity; and (b) major crops in America are on this basis highly vulnerable. In support of the latter, they cite data on the extent to which small numbers of varieties dominated crop acreage in the U.S. in 1969. These indicate that for 12 major crops, 50% or more of the total acreage was planted with 9 or fewer major varieties. For example, 71% of the corn acreage was planted to only six varieties and a single variety accounted for 69% of sweet-potato acreage.

If vulnerability stems from genetic uniformity, genetic diversity would appear to provide a solution to the problem. One approach to diversity is the use of multiline varieties in which a number of different genotypes are grown in mixed stands. Some success with this approach has been reported with small grains raised as feed for livestock (Browning and Frey, 1969). In crops raised for human consumption, it will probably be difficult to develop multilines that will meet the more stringent requirements of processors and consumers.

A second approach to diversity is the development of varieties with polygenic (general) resistance to replace those which have only one or a few genes (specific resistance). If a permanent high level of resistance could be achieved, this approach would be justified despite the difficulties imposed on plant breeding programs by the need to combine large numbers of genes in a single variety. However, general resistance usually does not reach the level conferred by specific resistance and there is no assurance that general resistance will remain stable. The theoretical and practical aspects of this approach have been considered in detail by Day (1974).

Recognition of the hazards involved in specific resistance does not justify the conclusion that the use of specific genes should be abandoned in breeding for disease resistance. Specific gene resistance has provided enduring protection against many diseases for which no other effective method of control is known. Fears have been widely expressed that specific resistance, through selective effects on pathogen populations, favors the development of new virulent races. Both the Victoria blight and southern corn leaf blight epidemics were brought under control by a sudden shift to varieties with specific resistance. Although this occurred in the presence of huge populations of the pathogen, new virulent races did not appear. It should be noted that both epidemics could have been avoided if early warning signs had been recognized. This suggests that increased vigilance and more extensive testing for new races can provide means for prevention of epidemics.

Chapter 6 Nature of the Physiological Syndrome

Throughout this book attention has been drawn to the similarities in physiological changes that have been observed in plants attacked by a wide variety of pathogenic agents. Although exceptions occur and only a small fraction of all diseases have been examined, the characteristic changes in structure, function, and metabolism discussed in Chapter 3 can be considered together as a physiological syndrome. So viewed, this physiological syndrome can serve as a common denominator for plant pathogenesis.

In the past, the tendency has been to concentrate on a single physiological change, rather than on the syndrome, as the crucial event in pathogenesis. Initially, respiration received most attention, because of its vital role in providing energy. Interest in respiration faded after the discovery that changes in permeability induced by victorin could be detected well in advance of changes in respiration (Chapter 3.B.I.3). Similar results with other pathotoxins, pathogen-produced pectic enzymes and pathogens themselves shifted the emphasis to permeability changes as crucial and possibly primary events in pathogenesis. Respiratory and other changes in metabolism were relegated to secondary roles by most investigators.

The dominance of membrane theory led many in the Western Hemisphere to equate early changes in permeability during pathogenesis with effects on the plasmalemma (Page, 1972). Despite extensive searches, especially with selective pathotoxins, direct evidence that initial changes in permeability result from effects on the plasmalemma has not been forthcoming. Instead, the initial response which occurs during the first five minutes after exposure to victorin, appears to reflect an effect on the cell wall (Fig. 15). If this interpretation is correct, changes in protoplast permeability induced by victorin, like those in respiration and others clearly linked to metabolism, occur only after a lag of 20–30 min.

Since pathogenic agents must penetrate the cell wall before they attack the protoplast, it is not surprising that the earliest host responses seen in electron micrographs, changes in staining (Figs. 4a, 5b), swelling (Fig. 3), modifications (Figs. 6a, 7a) and papillae (Figs. 2, 4a), are all associated with cell walls. What was surprising, at least to the writer, was evidence that the synthesis of cell-wall components increases in proportion to the increase in respiratory metabolism induced by victorin whereas synthesis of other cell fractions does not (Table 5). This and the ultrastructural changes seen in cell walls suggest that the cell has received a signal that the walls of its fortress are under attack. In response, the cell mobilizes all of its metabolic resources in an attempt to repair the damage.

The hypothesis that a signal originating in the cell wall triggers the physiological syndrome raises several questions. What kind of a signal could be generated by such diverse agents as pathotoxins, enzymes, viruses, and various pathogenic organisms? How is the signal sent? How is it received? One possible answer involves negative charges at cell surfaces which appear to be essential for the maintenance of normal cell permeability and function.

The importance of negative surface charges in the exclusion of toxic anions and the disastrous consequences which follow discharge of cell surfaces with dilute mineral acids have been discussed in detail by Lundegårdh (1966). Although their chemical nature is not known, these negatively charged surface materials must be fixed or semi-fixed since they are not readily removed by water or dilute salt solutions. Furthermore, they must be renewable otherwise root surfaces would be quickly discharged by cations in soil solutions. Lundegårdh (1966) points out that cellulose which is very weakly dissociated could not be expected to maintain a negative surface charge but that other cell-wall polysaccharides, especially pectins, could serve as fixed ion exchangers.

Support for the idea that polysaccharides or protein-polysaccharide complexes may function as negatively charged materials at cell surfaces has come from ultrastructural studies of slime secretion by epidermal and outer cap cells of roots. These cells differ in three ways from those in the root interior. Their walls stain very intensely after fixation in potassium permanganate, their endoplasmic reticulum is arranged in parallel profiles and their Golgi dictyosomes are hypersecretory. There is good evidence that densely stained Golgi vesicles secreted by these cells migrate out of the protoplast and contribute to the extruded slime which, at least in part, is made up of anionic polysaccharides (Morré and Mollenhauer, 1973). Such secretions, which are not readily soluble in water, would appear to be a likely source of fixed, renewable, negative surface charges. Since the cells involved are those in direct contact with the soil solution, it is possible that such secretions serve to recharge cell surfaces which are constantly being discharged by cations and that discharge at the cell surface triggers secretory activity.

Other than modifications in cell walls, the earliest changes in ultrastructure of interior cells of susceptible oat roots treated with victorin are an increase in staining of cell walls, the induction of hypersecretory activity by the Golgi apparatus and the formation of parallel profiles of endoplasmic reticulum (Hanchey et al., 1968). In other words, victorin transforms cells of the root interior into cells which resemble outer root cap and epidermal cells of untreated roots. Since at physiological pH victorin is highly cationic (Luke and Gracen, 1972), this would be expected if the characteristic features of outer cap and epidermal cells of normal roots are the result of their exposure to cations in the soil solution. It may be noted that increased Golgi secretory activity and the formation of parallel profiles of endoplasmic reticulum are characteristic of many diseased plant tissues (Chapter 3.A.II.2 and Bracker and Littlefield, 1973).

Secretory products of the Golgi apparatus have long been thought to play a role in cell-wall synthesis. In the endomembrane concept of Morré and Mollenhauer (1973) both the Golgi and the endoplasmic reticulum function in the secretion of materials from the cytoplasm through the plasmalemma and into the cell wall. Such secretory activity, especially the synthesis of polysaccharides or polysaccharide-protein complexes, would have a high energy requirement. If pathogenic agents through effects on surface charges trigger the synthesis and secretion of cell-wall components, the normal constraint of availability of phosphate acceptors would be removed and this would result in the rise in respiration which characterizes diseased tissues. Increased secretory activity would also account for the observation that increases in respiration induced by victorin are accompanied by increases in the incorporation of labeled sugars into cell-wall components (Table 5). Increased lignification, which is often observed in deseased tissues, requires incorporation of phenolic compounds into cell-wall components and this also may be linked to secretory activity.

The clearest evidence that the physiological syndrome may be triggered by a change in surface charges comes from measurements of the electrochemical potential of giant cells (syncytia) in oat roots treated with victorin. Such giant cells are packed with cytoplasm and do not have large vacuoles. Therefore the measured potential difference is across a layer of cytoplasm bounded by the plasmalemma and the cell wall. These measurements show a sharp initial discharge during the first five minutes after victorin is applied, then a slight recovery followed by a second prolonged discharge (Fig. 15c). If the data in Fig. 15 were interpreted correctly in Chapter 3.B.I.3. it is the second change which occurs 20–30 min after treatment with victorin that reflects a disruption of protoplast permeability. Since this is essentially the same lag required for changes in respiration and other processes linked to metabolism there is no basis for assigning primary or secondary roles to any of these effects.

On the basis of the proposed hypothesis, the 20–30 min lag before changes occur in metabolism or protoplast permeability in response to victorin represents the time required for a signal to be sent from the cell surface, received by an interior center and acted upon by the cell's metabolic machinery. This seems like a rather long time and suggests that the process or processes activated must have been in a quiescent state. Under normal conditions most plant cells do not exhibit secretory activity and the synthesis of secretory materials, formation of vesicles, vesicular migration and excretion, would require some time to be manifest. It is curious, but probably coincidental, that a number of other responses of plants occur with a lag time similar to that required for metabolic responses to victorin. Among these are responses to auxin (Saftner and Evans, 1974), the time required to wash loosely bound materials from cell-wall free space (Keck and Hodges, 1973) and to establish a steady, metabolically dependent rate of salt uptake (Lundegårdh, 1966).

The possibility that changes in negative surface charges precede changes in protoplast permeability and metabolism in diseases caused by agents other than pathotoxins has not been explored. Time-course studies of changes in electrochemical potentials in relation to permeability changes might shed new light on the nature of lethal effects of pectolytic enzymes which cause tissue maceration. It would also be of interest to know whether or not such enzymes induce secretory activity.

It is probably not necessary to emphasize that this chapter is even more speculative in nature than the preceding one. The hypothesis that a change in surface charge triggers the physiological syndrome is based largely on results with victorin, and a number of other hypotheses about the mode of action of this pathotoxin have been advanced (Luke and Gracen, 1972). Even the concept of a physiological syndrome as a sequence of changes in pathogenesis triggered by some initial event rests on assumptions unsupported by direct evidence. There may be no trigger and the sequence may represent unrelated or coincidental events (Wheeler and Hanchey, 1968). Whether the hypothesis is correct or not, it will have served its purpose if it stimulates fruitful new experimentation.

References

Abeles, F. B.: Ethylene in plant biology. London-New York: Academic Press 1973.

Adams, P. B., Sproston, T., Tietz, H., Major, R. T.: Studies on the disease resistance of Ginkgo biloba. Phytopathology **52**, 233–236 (1962).

Albersheim, P., Bauer, W., Keestra, K., Talmadge, K. W.: The structure of the wall of suspension-cultured sycamore cells. In: Loewus, F. (Ed.): Biogenesis of Plant Cell Wall Polysaccharides, pp. 117–147. London-New York: Academic Press 1973.

Albersheim, P., Jones, T. M., English, P. D.: Biochemistry of the cell wall in relation to infective processes. Ann. Rev. Phytopathology **7**, 171–194 (1969).

Allen, P. J.: Toxins and tissue respiration. Phytopathology **43**, 221–229 (1953).

Allison, A. V., Shalla, T. A.: The ultrastructure of local lesions induced by potato virus X: a sequence of cytological events in the course of infection. Phytopathology **64**, 784–793 (1974).

Arntzen, C. J., Haugh, M. F., Bobick, S.: Induction of stomatal closure by Helminthosporium maydis pathotoxin. Plant Physiol. **52**, 569–574 (1973).

Atkin, C. L., Neilands, J. B.: Leaf infections: siderochromes (natural polyhydroxamates) mimic the "green island" effect. Science **176**, 300–302 (1972).

Backman, P. A., DeVay, J. E.: Studies on the mode of action and biogenesis of the phytotoxin syringomycin. Physiol. Plant Pathology **1**, 215–233 (1971).

Bateman, D. F.: The polygalacturonase produced by Sclerotium rolfsii. Physiol. Plant Pathology **2**, 175–184 (1972).

Bateman, D. F., Daly, J. M.: The respiratory pattern of Rhizoctonia-infected hypocotyls in relation to lesion maturation. Phytopathology **57**, 127–131 (1967).

Bateman, D. F., Millar, R. L.: Pectic enzymes in tissue degradation. Ann. Rev. Phytopathology **4**, 119–146 (1966).

Beardsley, R. E.: The inception phase in the crown gall disease. Progr. Exp. Tumor Res. **15**, 1–75 (1972).

Beckman, C. H.: Host responses to vascular infection. Ann. Rev. Phytopathology **2**, 231–252 (1964).

Beevers, H.: Respiratory metabolism in plants. Evanston-White Plains: Row, Peterson and Co. 1961.

Bell, A. A.: Respiratory metabolism of Phaseolus vulgaris infected with alfalfa mosaic and southern bean mosaic viruses. Phytopathology **54**, 914–922 (1964).

Black, H. S., Wheeler, H.: Biochemical effects of victorin on oat tissues and mitochondria. Am. J. Botany **53**, 1008–1012 (1966).

Bracker, C. E., Littlefield, L. J.: Structural concepts of host-pathogen interfaces. In: Byrde, R. J. W., Cutting, C. V. (Eds.): Fungal Pathogenicity and the Plant's Response, pp. 159–318. London-New York: Academic Press 1973.

Brown, W.: Toxins and cell-wall dissolving enzymes in relation to plant disease. Ann. Rev. Phytopathology **3**, 1–18 (1965).

Browning, J. A., Frey, K. J.: Multiline cultivars as a means of disease control. Ann. Rev. Phytopathology **7**, 355–382 (1969).

Bushnell, W. R.: Symptom development in mildewed and rusted tissues. In: Mirocha, C. J., Uritani, I. (Eds.): The Dynamic Role of Molecular Constituents in Plant-Parasite Interaction, pp. 21–39. St. Paul: Am. Phytopathological Soc. 1967.

Byrde, R. J. W., Fielding, A. H.: Pectin methyl-trans-eliminase as the maceration factor of Sclerotinia fructigena and its significance in brown rot of apple. J. Gen. Microbiol. **52**, 287–297 (1968).

Campbell, R. N., Grogan, R. G.: Big-vein of lettuce and its transmission by Olpidium brassicae. Phytopathology **53**, 252–259 (1963).

Cocking, E. C.: Virus uptake, cell wall regeneration, and virus multiplication in isolated plant protoplasts. Intern. Rev. Cytol. **28**, 89–124 (1970).

Comstock, J. C., Martinson, C. A., Gengenbach, B. G.: Host specific of a toxin from Phyllosticta maydis for Texas cytoplasmically male-sterile maize. Phytopathology **63**, 1357–1361 (1973).

Cruickshank, I. A. M.: Phytoalexins. Ann. Rev. Phytopathology **1**, 351–374 (1963).

Daly, J. M.: Some metabolic consequences of infection by obligate parasites. In: Mirocha, C. J., Uritani, I. (Eds.): The Dynamic Role of Molecular Constituents in Plant-Parasite Interaction, pp. 144–164. St. Paul: Am. Phytopath. Soc. 1967.

Daly, J. M.: The use of near-isogenic lines in biochemical studies of the resistance of wheat to stem rust. Phytopathology **62**, 392–400 (1972).

Daly, J. M., Sayre, R. M.: Relations between growth and respiratory metabolism in safflower infected by Puccinia carthami. Phytopathology **47**, 163–168 (1957).

Davis, R. E., Whitcomb, R. F.: Mycoplasmas, Rickettsiae, and Chlamydiae: possible relation to yellows diseases and other disorders of plants and insects. Ann. Rev. Phytopathology **9**, 119–154 (1971).

Day, P. R.: Genetics of host-parasite interaction. San Francisco: W. H. Freeman and Co. 1974.

de Duve, C.: The lysosome in retrospect. In: Dingle, J. T., Fell, H. B. (Eds.): Lysosomes in Biology and Pathology: Frontiers of Biology, Vol. 1, pp. 3–37. Amsterdam-London: North Holland Publ. Co. 1969.

DeVay, J. E., Charudattan, R., Wimalajeewa, D. L. S.: Common antigenic determinants as a possible regulator of host-pathogen compatibility. Am. Naturalist **106**, 185–194 (1972).

Diener, T. O.: Viroids. Advances in Virus Res. **17**, 295–313 (1972).

Dimond, A. E.: Pathogenesis in the wilt diseases. Ann. Rev. Plant Physiol. **6**, 329–350 (1955).

Dimond, A. E.: Biophysics and biochemistry of the vascular wilt syndrome. Ann. Rev. Phytopathology **8**, 301–322 (1970).

Dimond, A. E., Waggoner, P. E.: The cause of epinastic symptoms in Fusarium wilt of tomatoes. Phytopathology **43**, 663–669 (1953).

Dixon, R. O.: Rhizobia (with particular reference to relationships with host plants). Ann. Rev. Microbiol. **23**, 137–158 (1969).

Drysdale, R. B., Langcake, P.: Response of tomato to infection by Fusarium oxysporum f. lycopersici. In: Byrde, R. J. W., Cutting, C. V. (Eds.): Fungal Pathogenicity and the Plant's Response, pp. 423–436. London-New York: Academic Press 1973.

Duniway, J. M.: Pathogen-induced changes in host water relations. Phytopathology **63**, 458–466 (1973).

Durbin, R. D.: Obligate parasites: effect on the movement of solutes and water. In: Mirocha, C. J., Uritani, I. (Eds.): The Dynamic Role of Molecular Constituents in Plant-Parasite Interaction, pp. 80–99. St. Paul: Am. Phytopathological Soc. 1967.

Durbin, R. D.: Bacterial phytotoxins: a survey of occurrence, mode of action and composition. In: Wood, R. K. S., Ballio, A., Graniti, A. (Eds.): Phytotoxins in Plant Diseases, pp. 19–33. London-New York: Academic Press 1972.

Dwurazna, M. M., Weintraub, M.: The respiratory pathways of tobacco leaves infected with potato virus X. Can. J. Botany **47**, 731–736 (1969).

Dyer, T. A., Scott, K. J.: Decrease in chloroplast content of barley leaves infected with powdery mildew. Nature **236**, 237–238 (1972).

Ehrlich, M. A., Ehrlich, H. G.: Fine structure of the host-parasite interfaces in mycoparasitism. Ann. Rev. Phytopathology **9**, 155–184 (1971).

Ellingboe, A. H.: Genetics and physiology of primary infection by Erysiphe graminis. Phytopathology **62**, 401–406 (1972).

Farkas, G. L., Király, A.: Studies on the respiration of wheat infected with stem rust and powdery mildew. Physiol. Plant. **8**, 877–887 (1955).

Favret, E. A.: The host-pathogen system and its genetic relationships. In: Robert A. Nilan (Ed.): Barley Genetics II., pp. 457–471. Pullman: Washington State University Press 1971.

Finlayson, G.R., Chrambach, A.: Isoelectric focusing in polyacrylamide gel and its preparative application. Analyt. Biochem. **40**, 292–311 (1971).
Flor, H.H.: Host-parasite interaction in flax rust: its genetics and other implications. Phytopathology **45**, 680–685 (1955).
Flor, H.H.: Current status of the gene-for-gene concept. Ann. Rev. Phytopathology **9**, 275–296 (1971).
Garibaldi, H., Bateman, D.F.: Pectic enzymes produced by Erwinia chrysanthemi and their effects on plant tissue. Physiol. Plant Pathology **1**, 25–40 (1971).
Garrett, S.D.: Biology of root-infecting fungi. Cambridge: University Press 1956.
Gäumann, E.: Principles of plant infection (English translation by W.B. Brierley). London: Crosley Lockwood and Son 1955.
Gäumann, E.: The mechanisms of fusaric acid injury. Phytopathology **48**, 670–686 (1958).
Ghabrial, S.A., Pirone, T.P.: Physiology of tobacco etch virus-induced wilt of Tabasco peppers. Virology **31**, 154–162 (1967).
Goodman, R.N.: Protection of apple stem tissue against Erwinia amylovora infection by avirulent strains and three other bacterial species. Phytopathology **57**, 22–24 (1967).
Goodman, R.N.: Electrolyte leakage and membrane damage in relation to bacterial population, pH, and ammonia production in tobacco leaf tissue inoculated with Pseudomonas pisi. Phytopathology **62**, 1327–1331 (1972).
Goodman, R.N., Huang, J.S., Huang, Pi-Yu.: Host-specific phytotoxic polysaccharide from apple tissue infected by Erwinia amylovora. Science **183**, 1081–1082 (1974).
Goodman, R.N., Király, Z., Zaitlin, M.: The biochemistry and physiology of infectious plant disease. Princeton: D. van Nostrand Company 1967.
Graniti, A.: The evolution of the toxin concept in plant pathology. In: Wood, R.K.S., Ballio, A., Graniti, A. (Eds.): Phytotoxins in Plant Diseases, pp. 1–18. London-New York: Academic Press 1972.
Grover, R.K.: Participation of host exudate chemicals in appressorium formation by Colletotrichum piperatum. In: Preece, T.F., Dickinson, C.H. (Eds.): Ecology of Leaf Surface Microorganisms, pp. 509–518. London-New York: Academic Press 1971.
Hadwiger, L.A., Schwochau, M.E.: Host resistance responses—an induction hypothesis. Phytopathology **59**, 223–227 (1969).
Hall, J.A., Wood, R.K.S.: The killing of plant cells by pectolytic enzymes. In: Byrde, R.J.W., Cutting, C.V. (Eds.): Fungal Pathogenicity and the Plant's Response, pp. 19–38. London-New York: Academic Press 1973.
Hall, R.: Pathogenism and parasitism as concepts of symbiotic relationships. Phytopathology **64**, 576–577 (1974).
Hampton, R.E.: An oxidation product of chlorogenic acid in tobacco leaves infected with tobacco streak virus. Phytopathology **60**, 1677–1681 (1970).
Hanchey, P., Pastalka, T., Novacky, A.: Ultrastructure of tobacco protected against the hypersensitive response. Abstr. 243. Vancouver: 66 Annual Meeting of the Amer. Phytopathological Soc. 1974.
Hanchey, P., Wheeler, H.: Pathological changes in ultrastructure; false plasmolysis. Can. J. Botany **47**, 675–678 (1969).
Hanchey, P., Wheeler, H., Luke, H.H.: Pathological changes in ultrastructure: effects of victorin on oat roots. Am. J. Botany **55**, 53–61 (1968).
Heitefuss, R.: Nucleic acid metabolism in obligate parasitism. Ann. Rev. Phytopathology **4**, 221–244 (1966).
Heitefuss, R., Fuchs, W.H.: Phosphatstoffwechsel und Sauerstoffaufnahme in Weizenkeimpflanzen nach Infektion mit Puccinia graminis tritici. Phytopathol. Z. **46**, 174–198 (1963).
Henry, S.M.: Foreword. In: Henry, S.M. (Ed.): Symbiosis, Vol. I, Associations of Microorganisms, Plants, and Marine Organisms, pp. ix–xi. London-New York: Academic Press 1966.
Hess, W.M.: Ultrastructure of onion roots infected with Pyrenochaeta terrestris, a fungus parasite. Am. J. Botany **56**, 832–845 (1969).
Hewitt, W.B., Raski, D.J., Goheen, A.C.: Nematode vector of soil-borne fanleaf virus of grapevines. Phytopathology **48**, 586–595 (1958).

Hickman, C.J., Ho, H.H.: Behaviour of zoospores in plant-pathogenic Phycomycetes. Ann. Rev. Phytopathology **4**, 195–220 (1966).

Higgins, V.J., Millar, R.L.: Comparative abilities of Stemphylium botryosum and Helminthosporium turcicum to induce and degrade a phytoalexin from alfalfa. Phytopathology **59**, 1493–1499 (1969).

Hislop, E.C., Hoad, G.V., Archer, S.A.: The involvement of ethylene in plant diseases. In: Byrde, R.J.W., Cutting, C.V. (Eds.): Fungal Pathogenicity and the Plant's Response, pp. 87–117. London-New York: Academic Press 1973.

Hooker, A.L.: Southern leaf blight of corn-present status and future prospects. J. Environm. Quality **1**, 244–249 (1972).

Ingham, J.: Phytoalexins and other natural products as factors in plant disease resistance. Botan. Rev. **38**, 343–424 (1972).

Kaars Sijpesteijn, A., Van Dijkman, A.: The host-parasite interactions in resistance of tomatoes to Cladosporium fulvum. In: Byrde, R.J.W., Cutting, C.V. (Eds.): Fungal Pathogenicity and the Plant's Response, pp. 437–448. London-New York: Academic Press 1973.

Kajiwara, T.: Structure and physiology of haustoria of various parasites. In: Akai, S., Ouchi, S. (Eds.): Morphological and Biochemical Events in Plant-Parasite Interaction, pp. 255–277. Tokyo: Phytopathological Soc. Japan 1971.

Keck, R.W., Hodges, T.K.: Membrane permeability in plants: changes induced by host-specific pathotoxins. Phytopathology **63**, 226–230 (1973).

Keen, N.T., Horsch, R.: Hydroxyphaseollin production by various soybean tissues: a warning against use of "unnatural" host-parasite systems. Phytopathology **62**, 439–442 (1972).

Kern, H.: Phytotoxins produced by Fusaria. In: Wood, R.K.S., Ballio, A., Graniti, A. (Eds.): Phytotoxins in Plant Diseases, pp. 35–48. London-New York: Academic Press 1972.

Keskin, B., Fuchs, W.H.: Der Infektionsvorgang bei Polymyxa betae. Arch. Mikrobiol. **68**, 218–226 (1969).

Khew, K.L., Zentmyer, G.A.: Electrotactic response of zoospores of seven species of Phytophthora. Phytopathology **64**, 500–507 (1974).

Kosuge, T.: The role of phenolics in host response to infection. Ann. Rev. Phytopathology **7**, 195–222 (1969).

Kraft, J.M., Endo, R.M., Erwin, D.C.: Infection of primary roots of bentgrass by zoospores of Pythium aphanidermatum. Phytopathology **57**, 86–90 (1967).

Krupka, L.R.: Metabolism of oats susceptible to Helminthosporium victoriae and victorin. Phytopathology **49**, 587–594 (1959).

Kuć, J.: Phytoalexins. Ann. Rev. Phytopathology **10**, 204–232 (1972).

Lamport, D.T.A.: The glycopeptide linkages of extensin: O-D-galactosyl and O-L-arabinosyl hydroxyproline. In: Loewus, F. (Ed.): Biogenesis of Plant Cell Wall Polysaccharides, pp. 149–164. London-New York: Academic Press 1973.

Levine, R.P.: Genetic dissection of photosynthesis. Science **162**, 768–771 (1968).

Littlefield, L.J.: Flax rust resistance induced by prior inoculation with an avirulent race of Melampsora lini. Phytopathology **59**, 1323–1328 (1969).

Litzenberger, S.C.: Nature of susceptibility to Helminthosporium victoriae and resistance to Puccinia coronata in Victoria oats. Phytopathology **39**, 300–318 (1949).

Loebenstein, G.: Localization and induced resistance in virus-infected plants. Ann. Rev. Phytopathology **10**, 177–206 (1972).

Loegering, W.Q.: Specificity in plant disease. In: Biology of Rust Resistance in Forest Trees, pp. 29–41. U.S. Dept. Agriculture Misc. Pub. No. 1221, 1972.

Loegering, W.Q., Powers, H.R., Jr.: Inheritance of pathogenicity in a cross of physiological races 111 and 36 of Puccinia graminis f. sp. tritici. Phytopathology **52**, 547–554 (1962).

Luke, H.H., Freeman, T.E.: Effects of victorin on Krebs cycle intermediates of a susceptible oat variety. Phytopathology **55**, 967–969 (1965).

Luke, H.H., Gracen, V.E., Jr.: Helminthosporium toxins. In: Kadis, S., Ciegler, A., Ajl, S.J. (Eds.): Microbial Toxins, Vol. 8 Fungal Toxins, pp. 139–168. London-New York: Academic Press 1972.

Luke, H.H., Wallace, A.T.: Sensitivity of induced mutants of an Avena cultivar to victorin at different temperatures. Phytopathology **59**, 1769–1770 (1969).

Luke, H. H., Wheeler, H. E.: Toxin production by Helminthosporium victoriae. Phytopathology **45**, 453–458 (1955).

Luke, H. H., Wheeler, H. E., Wallace, A. T.: Victoria-type resistance to crown-rust separated from susceptibility to Helminthosporium blight in oats. Phytopathology **50**, 205–209 (1960).

Lund, B. M.: The effect of certain bacteria on ethylene production by plant tissue. In: Byrde, R. J. W., Cutting, C. V. (Eds.): Fungal Pathogenicity and the Plant's Response, pp. 69–86. London-New York: Academic Press 1973.

Lundegardh, H.: In: James, W. O. (Ed.): Plant physiology (English translation by F. M. Irvine. New York: Am. Elsevier 1966.

Luttrell, E. S.: Parasitism of fungi on vascular plants. Mycologia **66**, 1–15 (1974).

MacDonald, P. W., Strobel, G. A.: Adenosine diphosphate-glucose pyrophosphorylase control of starch accumulation in rust-infected wheat leaves. Plant Physiol. **46**, 126–135 (1970).

Mansfield, J. W., Deverall, B. J.: Mode of action of pollen in breaking resistance of Vicia faba to Botrytis cinerea. Nature **232**, 339 (1971).

Martin, J. T.: Role of cuticle in the defense against plant disease. Ann. Rev. Phytopathology **2**, 81–100 (1964).

Matta, A.: Microbial penetration and immunization of uncongenial host plants. Ann. Rev. Phytopathology **9**, 387–410 (1971).

Matthews, R. E. F.: Plant virology. London-New York: Academic Press 1970.

McIntyre, J. L.: Protection of pear against fire blight by bacteria, bacterial sonicates and bacterial deoxyribonucleic acid. Ph.D. Dissertation. Purdue Univ. Library (1974).

McKeen, W. E.: Mode of penetration of epidermal cell walls of Vicia faba by Botrytis cinerea. Phytopathology **64**, 461–467 (1974).

Meehan, F., Murphy, H. C.: Differential phytotoxicity of metabolic by-products of Helminthosporium victoriae. Science **106**, 270–271 (1947).

Mercer, P. C., Wood, R. K. S., Greenwood, A. D.: Initial infection of Phaseolus vulgaris by Colletotrichum lindemuthianum. In: Preece, T. F., Dickinson, C. H. (Eds.): Ecology of Leaf Surface Micro-organisms, pp. 381–389. London-New York: Academic Press 1971.

Metlitskii, L. V., Ozeretskovskaya, O. L.: Plant immunity: biochemical aspects of plant resistance to parasitic fungi. New York: Plenum Press 1968.

Miller, R. J., Koeppe, D. E.: Southern corn leaf blight: susceptible and resistant mitochondria. Science **173**, 67–69 (1971).

Mirocha, C. J.: Phytotoxins and metabolism. In: Wood, R. K. S., Ballio, A., Graniti, A. (Eds.): Phytotoxins in Plant Diseases, pp. 191–209. London-New York: Academic Press 1972.

Mirocha, C. J., Rick, P. D.: Carbon dioxide fixation in the dark as a nutritional factor in parasitism. In: Mirocha, C. J., Uritani, I. (Eds.): The Dynamic Role of Molecular Constituents in Plant-Parasite Interaction, pp. 121–143. St. Paul: Am. Phytopathological Soc. 1967.

Mirocha, C. J., Zaki, A. I.: Fluctuation in amount of starch in host plants invaded by rust and mildew fungi. Phytopathology **56**, 1220–1224 (1966).

Mori, R.: Studies on the resistance of Japanese pears to the black spot disease. The Lib. Arts. J., Tottori Univ. **13**, 53–119 (1962).

Morré, D. J., Mollenhauer, H. H.: The endomembrane concept: A functional integration of endoplasmic reticulum and Golgi apparatus. In: Robards, A. W. (Ed.): Dynamics of Plant Ultrastructure, pp. 84–131. New York: McGraw-Hill 1973.

Mount, M. S., Bateman, D. F., Basham, H. G.: Induction of electrolyte loss, tissue maceration, and cellular death of potato tissue by an endopolygalacturonate trans-eliminase. Phytopathology **60**, 924–931 (1970).

Mullen, J. M., Bateman, D. F.: Production of an endopolygalacturonate trans-eliminase by a potato dry-rot pathogen, Fusarium roseum 'Avenaceum,' in culture and in diseased tissue. Physiol. Plant Path. **1**, 363–373 (1971).

Müller, K. O., Börger, H.: Experimentelle Untersuchungen über die Phytophthora: Resistenz der Kartoffel. Arb. Biol. Abt. (Aust. Reichsanst.) Berlin **23**, 183–231 (1940).

National Academy of Sciences (U.S.A.): Genetic vulnerability of major crops. Washington, D.C. 1972.

Neish, A. C.: Major pathways of biosynthesis of phenols. In: Harborne, J. B. (Ed.): Biochemistry of Phenolic Compounds, pp. 295–359. London-New York: Academic Press 1964.

Nelson, R. R.: The meaning of disease resistance in plants. In: Nelson, R. R. (Ed.): Breeding Plants for Disease Resistance: Concepts and Applications, pp. 13–25. University Park-London: Pennsylvania State University Press 1973.

Neukom, H.: Über den Abbau von Pektinstoffen. Schweiz. Landw. Forsch. **2**, 112–122 (1963).

Nicholson, R. L., Kuć, J., Williams, E. B.: Histochemical demonstration of transitory esterase activity in Venturia inaequalis. Phytopathology **62**, 1242–1247 (1972).

Novacky, A., Hampton, R. E.: Peroxidase isozymes in virus-infected plants. Phytopathology **58**, 301–305 (1968).

Otani, H., Nishimura, S., Kohmoto, K.: Nature of specific susceptibility to Alternaria kikuchiana in Nijisseiki cultivar among Japanese pears (III) Chemical and thermal protection against effect of host-specific toxin. Ann. Phytopath. Soc. Japan **40**, 59–66 (1974).

Paddock, W. C.: Histological study of suscept-pathogen relationships between Helminthosporium victoriae and seedling oat leaves. Cornell Agr. Expt. Sta. Memoir 315 (1953).

Page, O. T.: Effect of phytotoxins on the permeability of cell membranes. In: Wood, R. K. S., Ballio, A., Graniti, A. (Eds.): Phytotoxins in Plant Diseases, pp. 211–225. London-New York: Academic Press 1972.

Pappelis, A. J., Kulfinski, F. B.: Quantitative interferometry of epidermal nuclei in senescing and infected leaves. In: Preece, T. F., Dickinson, C. H. (Eds.): Ecology of Leaf Surface Micro-organisms, pp. 353–366. London-New York: Academic Press 1971.

Patil, S. S.: Toxins produced by phytopathogenic bacteria. Ann. Rev. Phytopathology **12**, 259–279 (1974).

Patrick, Z. A., Toussoun, T. A., Koch, L. W.: Effect of crop-residue decomposition products on plant roots. Ann. Rev. Phytopathology **2**, 267–292 (1964).

Pitt, D., Galpin, M.: Role of lysosomal enzymes in pathogenicity. In: Byrde, R. J. W., Cutting, C. V. (Eds.): Fungal Pathogenicity and the Plant's Response, pp. 449–467. London-New York: Academic Press 1973.

Politis, D. J., Wheeler, H.: Ultrastructural study of penetration of maize leaves by Colletotrichum graminicola. Physiol. Plant Pathol. **3**, 465–471 (1973).

Pozsár, B. I., Király, Z.: Effect of rust infection on oxidative phosphorylation of wheat leaves. Nature **182**, 1686–1687 (1958).

Pozsár, B. I., Király, Z.: Phloem transport in rust infected plants and the cytokinin-directed long-distance movement of nutrients. Phytopathol. Z. **56**, 297–309 (1966).

Preece, T. F., Dickinson, C. H. (Eds.): Ecology of Leaf Surface Micro-organisms. London-New York: Academic Press 1971.

Pringle, R. B.: Chemistry of host-specific phytotoxins. In: Wood, R. K. S., Ballio, A., Graniti, A. (Eds.): Phytotoxins in Plant Diseases, pp. 139–155. London-New York: Academic Press 1972.

Pueppke, S. G., Van Etten, H. D.: Pisatin accumulation and lesion development in peas infected with Aphanomyces euteiches, Fusarium solani f. sp. pisi, or Rhizoctonia solani. Phytopathology **64**, 1433–1440 (1974).

Rawn, C. D.: Victorin-induced changes in carbohydrate metabolism in oat leaves. Ph.D. dissertation, Univ. of Kentucky Library (1974).

Retig, N., Chet, I.: Catechol-induced resistance of tomato plants to Fusarium wilt. Physiol. Plant Pathology **4**, 469–475 (1974).

Rubin, B. A., Artsikhovskaya, E. V.: Biochemistry and physiology of plant immunity. Oxford: Pergamon Press 1963.

Rubin, B. A., Artsikhovskaya, E. V.: Biochemistry of pathological darkening of plant tissues. Ann. Rev. Phytopathology **2**, 157–178 (1964).

Saftner, R. A., Evans, M. L.: Selective effects of victorin on growth and the auxin response in Avena. Plant Physiol. **53**, 382–387 (1974).

Samaddar, K. R., Scheffer, R. P.: Effect of the specific toxin in Helminthosporium victoriae on host cell membranes. Plant Physiol. **43**, 21–28 (1968).

Schafer, J. F.: Tolerance to plant disease. Ann. Rev. Phytopathology **9**, 235–252 (1971).

Scheffer, R.P., Yoder, O.C.: Host-specific toxins and selective toxicity. In: Wood, R.K.S., Ballio, A., Graniti, A. (Eds.): Phytotoxins in Plant Diseases, pp.251–272. London-New York: Academic Press 1972.
Sempio, C.: Metabolic resistance to plant diseases. Phytopathology **40**, 799–819 (1950).
Sempio, C.: The host is starved. In: Horsfall, J.G., Dimond, A.E. (Eds.): Plant Pathology: An Advanced Treatise, Vol. I, The Diseased Plant, pp.277–312. London-New York: Academic Press 1959.
Sequeira, L.: Hormone metabolism in diseased plants. Ann. Rev. Plant Physiol. **24**, 353–380 (1973).
Shaw, M., Samborski, D.J.: The physiology of host-parasite relations. I. The accumulation of radioactive substances at infections of facultative and obligate parasites including tobacco mosaic virus. Can. J. Botany **34**, 389–405 (1956).
Shaw, M., Samborski, D.J.: The physiology of host-parasite relations. III. The pattern of respiration in rusted and mildewed cereal leaves. Can. J. Botany **35**, 389–407 (1957).
Slykhuis, J.T.: Mites as vectors of plant viruses. In: Maramorosch, K. (Ed.): Viruses, Vectors and Vegetation, pp.121–141. New York: Interscience Publ. 1969.
Smedegard-Peterson, V., Nelson, R.R.: The production of a host-specific pathotoxin by Cochliobolus heterostrophus. Can. J. Botany **47**, 951–957 (1969).
Spalding, D.H., Abdul-Baki, A.A.: In vitro and in vivo production of pectin lyase by Penicillium expansum. Phytopathology **63**, 231–235 (1973).
Stahmann, M.A., Demorest, D.M.: Changes in enzymes of host and pathogen with special reference to peroxidase interaction. In: Byrde, R.J.W., Cutting, C.V. (Eds.): Fungal Pathogenicity and the Plant's Response, pp.405–422. London-New York: Academic Press 1973.
Staub, T., Dahmen, H., Schwinn, F.J.: Light- and scanning-electron microscopy of cucumber and barley powdery mildew on host and nonhost plants. Phytopathology **64**, 364–372 (1974).
Steiner, G.W., Byther, R.S.: Partial characterization and use of a host-specific toxin from *Helminthosporium sacchari* on sugarcane. Phytopathology **61**, 691–695 (1971).
Stoessl, A.: Antifungal compounds produced by higher plants. Recent Advan. Phytochem. **3**, 143–180 (1970).
Strobel, G.A.: Phytotoxins produced by plant parasites. Ann. Rev. Plant Physiol. 25, 541–566 (1974).
Swinburne, T.R.: The resistance of immature Bramley's seedling apples to rotting by Nectria galligena Bres. In: Byrde, R.J.W., Cutting, C.V. (Eds.): Fungal Pathogenicity and the Plant's Response, pp.365–382. London-New York: Academic Press 1973.
Takebe, I., Otsuki, Y.: Infection of tobacco mesophyll protoplasts by tobacco mosaic virus. Pro. Nat. Acad. Sci. U.S. **64**, 843–848 (1969).
Temmink, J.H.M., Campbell, R.N.: The ultrastructure of Olpidium brassicae. III. Infection of host roots. Can. J. Botany **47**, 421–424 (1969).
Templeton, G.E.: Alternaria toxins related to pathogenesis in plants. In: Kadis, S., Ciegler, A., Ajl, S.J. (Eds.): Microbial Toxins, Vol. VIII, Fungal Toxins, pp.169–192. London-New York: Academic Press 1972.
Thrower, L.B.: Terminology for plant parasites. Phytopathol. Z. **56**, 258–259 (1966).
Tomiyama, K.: Cytological and biochemical studies of the hypersensitive reaction of potato cells to Phytophthora infestans. In: Akai, S., Ouchi, S. (Eds.): Morphological and Biochemical Events in Plant-Parasite Interaction, pp.387–401. Tokyo: Phytopathological Soc. Japan 1971.
Tseng, T.C., Mount, M.S.: Toxicity of endopolygalacturonate trans-eliminase, phosphatidase, and protease to potato and cucumber tissue. Phytopathology **64**, 229–236 (1974).
Turner, N.C.: Stomatal behavior of Avena sativa treated with two phytotoxins, victorin and fusicoccin. Am. J. Botany **59**, 133–136 (1972).
Turner, N.C., Graniti, A.: Fusicoccin: a fungal toxin that opens stomata. Nature **223**, 1070–1071 (1969).
Uritani, I.: Protein changes in diseased plants. Ann. Rev. Phytopathology **9**, 211–234 (1971).
Van Andel, O.M., Fuchs, A.: Interference with plant growth regulation by microbial metabolites. In: Wood, R.K.S., Ballio, A., Graniti, A. (Eds.): Phytotoxins in Plant Diseases, pp.227–249. London-New York: Academic Press 1972.

Vesterberg, O., Svensson, H.: Isoelectric fractionation, analysis, and characterization of ampholytes in natural pH gradients. Acta. Chem. Scand. **20**, 820–834 (1966).

Von Broembsen, S. L., Hadwiger, L. A.: Characterization of disease resistance responses in certain gene-for-gene interactions between flax and Melampsora lini. Physiol. Plant Pathol. **2**, 207–215 (1972).

Wacek, T. J., Sequeira, L.: The peptidoglycan of Pseudomonas solanacearum: chemical composition and biological activity in relation to the hypersensitive reaction in tobacco. Physiol. Plant. Pathol. **3**, 363–369 (1973).

Wang, M. C., Keen, N. T.: Purification and characterization of endopolygalacturonase from Verticillium albo-atrum. Arch. Biochem. Biophys. **141**, 749–757 (1970).

Webster, J. M.: The host-parasite relationships of plant-parasitic nematodes. Advan. Parasitology **7**, 1–40 (1969).

Wells, H. D., Bell, D. K., Jaworski, C. A.: Efficacy of Trichoderma harzianum as a biocontrol for Sclerotium rolfsii. Phytopathology **62**, 442–447 (1972).

Wheeler, H.: Genetics of pathogenesis. In: Proceedings, Symposium on Potentials in Crop Protection, pp. 9–13. New York: State Agricultural Experiment Station, Geneva, Cornell University 1969.

Wheeler, H.: Cell wall and plasmalemma modifications in diseased and injured plant tissues. Can. J. Botany **52**, 1005–1009 (1974).

Wheeler, H., Hanchey, P. J.: Respiratory control: loss in mitochondria from diseased plants. Science **154**, 1569–1571 (1966).

Wheeler, H., Hanchey, P.: Permeability phenomena in plant disease. Ann. Rev. Phytopathology **6**, 331–350 (1968).

Wheeler, H., Luke, H. H.: Microbial toxins in plant disease. Ann. Rev. Microbiol. **17**, 223–242 (1963).

White, J. C., Horn, N. L.: The histology of Tabasco peppers infected with tobacco etch virus. Phytopathology **55**, 267–269 (1965).

Whitney, H. S., Shaw, M., Naylor, J. M.: The physiology of host-parasite relations. XII. A cytophotometric study of the distribution of DNA and RNA in rust-infected leaves. Can. J. Botany **40**, 1533–1544 (1962).

Williams, P. H., Aist, J. R., Bhattacharya, P. K.: Host-parasite relations in cabbage clubroot. In: Byrde, R. J. W., Cutting, C. V. (Eds.): Fungal Pathogenicity and the Plant's Response, pp. 141–158. London-New York: Academic Press 1973.

Wilson, C. L.: A lysosomal concept for plant pathology. Ann. Rev. Phytopathology **11**, 247–272 (1973).

Wood, R. K. S.: Physiological plant pathology. Blackwell Scientific Publications. Oxford-Edinburgh: 1967.

Wood, R. K. S.: Specificity in plant diseases. In: Byrde, R. J. W., Cutting, C. V. (Eds.): Fungal Pathogenicity and the Plant's Response, pp. 1–16. London-New York: Academic Press 1973.

Wood, R. K. S., Ballio, A., Graniti, A. (Eds.): Phytotoxins in Plant Diseases. London-New York: Academic Press 1972.

Wynn, W. K.: Personal communication (1974).

Yoder, O. C.: A selective toxin produced by Phyllosticta maydis. Phytopathology **63**, 1361–1366 (1973).

Subject Index

Encyclopedia of Plant Physiology, New Series

Editors: A. Pirson, M.H. Zimmermann

Volume 1
Transport in Plants I
Phloem Transport
Editors: M.H. Zimmermann, J.A. Milburn
With contributions by M.J. Canny, J. Dainty, A.F.G. Dixon, W. Eschrich, D.S. Fensom, D.R. Geiger, W. Heyser, W. Höll, J.A. Milburn, T.R.F. Nonweiler, M.V. Parthasarathy, J.S. Pate, A.J. Peel, S. Sovonick, D.C. Spanner, P.M.L. Tammes, M.T. Tyree, J. Van Die, H. Ziegler, M.H. Zimmermann

93 figures. Approx. 550 pages. 1975
ISBN 3-540-07314-0 Cloth DM 158,–
ISBN 0-387-07314-0 (North America) Cloth $68.00
Distribution rights for India: The Universal Book Stall, New Delhi

Prices are subject to change without notice

This book, as the first of three on transport of plants opening the new edition of the well-known Encyclopedia of Plant Physiology, is a carefully selected coverage of the study of phloem transport. For the first time, the conflicting theories to date are brought together in one informatory volume, parts I, II and III giving the factual evidence of the phenomenon of phloem transport, parts IV and V expounding the possible mechanisms, be they electro-osmotic flow, pressure flow, protoplasmic streaming, investigation with electron micrograph etc. The book represents the range of present-day research. The varied interpretations of 20 experts are skillfully anchored to the historical background of research on phloem transport, and a full account of the basic facts available for the necessary future research.

Springer-Verlag
Berlin
Heidelberg
New York

Advanced Series in Agricultural Sciences

Co-ordinating Editor: B. Yaron
Editors: G.W. Thomas, B.R. Sabey, Y. Vaadia, L.D. Van Vleck

Volume 1: A.P.A. VINK
Land Use in Advancing Agriculture
94 figures. 115 tables. X, 394 pages. 1975
ISBN 3-540-07091-5 Cloth DM 60,–
ISBN 0-387-07091-5 (North America) Cloth $25.80
Distribution rights for India: Allied Publishers Private Ltd., New Delhi

Prices are subject to change without notice

Contents: Land Use Surveys. – Land Utilization Types. – Land Resources. – Landscape Ecology and Land Conditions. – Land Evaluation. – Development of Land Use in Advancing Agriculture. – References. – Subject Index.

Agriculture is continuously advancing toward the economic and social goals demanded by human society. To achieve this advance, old and new technologies are being applied to various kinds of land resources. A careful study of land use and of different types of land resources involves a variety of methods for research and evaluation. Land management offers a means for maintaining a sound ecologic balance while meeting the demands of society. For optimal success, land use and land management have to be well adapted to both land resources and ecologic conditions. Land improvement is one way of adapting land resources to human requirements.

The book is addressed to agronomists, soil scientists, geographers, and planners, and to postgraduate students in these sciences. Its aim is to promote sound methods for land use that will meet human needs without infringing ecological principles.

Springer-Verlag
Berlin
Heidelberg
New York